Lecture Notes in Physics

Monographs

Springer
Berlin
Heidelberg
New York
Barcelona
Hong Kong
London
Milan
Paris
Tokyo

The Editorial Policy for Monographs

The series Lecture Notes in Physics reports new developments in physical research and teaching - quickly, informally, and at a high level. The type of material considered for publication in the monograph Series includes monographs presenting original research or new angles in a classical field. The timeliness of a manuscript is more important than its form, which may be preliminary or tentative. Manuscripts should be reasonably self-contained. They will often present not only results of the author(s) but also related work by other people and will provide sufficient motivation, examples, and applications.
The manuscripts or a detailed description thereof should be submitted either to one of the series editors or to the managing editor. The proposal is then carefully refereed. A final decision concerning publication can often only be made on the basis of the complete manuscript, but otherwise the editors will try to make a preliminary decision as definite as they can on the basis of the available information.
Manuscripts should be no less than 100 and preferably no more than 400 pages in length. Final manuscripts should be in English. They should include a table of contents and an informative introduction accessible also to readers not particularly familiar with the topic treated. Authors are free to use the material in other publications. However, if extensive use is made elsewhere, the publisher should be informed. Authors receive jointly 30 complimentary copies of their book. They are entitled to purchase further copies of their book at a reduced rate. No reprints of individual contributions can be supplied. No royalty is paid on Lecture Notes in Physics volumes. Commitment to publish is made by letter of interest rather than by signing a formal contract. Springer-Verlag secures the copyright for each volume.

The Production Process

The books are hardbound, and quality paper appropriate to the needs of the author(s) is used. Publication time is about ten weeks. More than twenty years of experience guarantee authors the best possible service. To reach the goal of rapid publication at a low price the technique of photographic reproduction from a camera-ready manuscript was chosen. This process shifts the main responsibility for the technical quality considerably from the publisher to the author. We therefore urge all authors to observe very carefully our guidelines for the preparation of camera-ready manuscripts, which we will supply on request. This applies especially to the quality of figures and halftones submitted for publication. Figures should be submitted as originals or glossy prints, as very often Xerox copies are not suitable for reproduction. For the same reason, any writing within figures should not be smaller than 2.5 mm. It might be useful to look at some of the volumes already published or, especially if some atypical text is planned, to write to the Physics Editorial Department of Springer-Verlag direct. This avoids mistakes and time-consuming correspondence during the production period.
As a special service, we offer free of charge LaTeX and TeX macro packages to format the text according to Springer-Verlag's quality requirements. We strongly recommend authors to make use of this offer, as the result will be a book of considerably improved technical quality.
For further information please contact Springer-Verlag, Physics Editorial Department II, Tiergartenstrasse 17, D-69121 Heidelberg, Germany.

Series homepage – http://www.springer.de/phys/books/lnpm

Wilfried Staude

Laser-Strophometry

High-Resolution Techniques for Velocity Gradient Measurements in Fluid Flows

Springer

Author

Wilfried Staude
Universität Bremen
Institut für Experimentelle Physik
28334 Bremen, Germany

Library of Congress Cataloging-in-Publication Data applied for.

Die Deutsche Bibliothek - CIP-Einheitsaufnahme

Staude, Wilfried:
Laser strophometry : high resolution techniques for velocity gradient measurements in fluid flows / Wilfried Staude. - Berlin ; Heidelberg ; New York ; Barcelona ; Hong Kong ; London ; Milan ; Paris ; Singapore ; Tokyo : Springer, 2001
(Lecture notes in physics : N.s. M, Monographs ; 69)
(Physics and astronomy online library)
ISBN 3-540-42622-1

ISSN 0940-7677 (Lecture Notes in Physics. Monographs)
ISBN 3-540-42622-1 Springer-Verlag Berlin Heidelberg New York

Springer-Verlag Berlin Heidelberg New York
a member of BertelsmannSpringer Science+Business Media GmbH

http://www.springer.de

Typesetting: Camera-ready by the author
Cover design: *design & production*, Heidelberg

Printed on acid-free paper
SPIN: 10851102 55/3141/du - 5 4 3 2 1 0

To Christa

Preface

This book is the result of two decades of research work which started with an accidental observation. One of my students, Dipl. phys. Volkmar Lenz, noticed that the speckle pattern of laser light scattered by a cuvette containing diluted milk performed a strange motion every time he came near the cuvette with his thumb. After thinking about this effect we came to the conclusion that this motion can only be caused by scattering particles with different velocities, as in the case of the diffraction pattern of an optical grating: A linear motion of the grating does not change the pattern whereas a rotation of the grating does. The observed speckle motion could then be explained qualitatively as produced by the inhomogeneous velocity of the convection within the cuvette which was produced by the heat of the thumb.

The theoretical treatment of this effect revealed that the velocity gradient of the light scattering medium is responsible for the speckle motion. The idea to use this effect for developing measurement techniques for velocity gradients arose almost immediately. For that purpose we had to develop not only experimental set-ups to measure the pattern velocity but also the theory which describes the connection between this velocity and the velocity gradient. The result of this work together with the description of a method developed by another group forms the contents of this book.

I am indebted to the students who worked in my laboratory and developed the measurement techniques. These were, in temporal order, Dr. Christoph Keveloh, Dr. Ulrich Schmidt, Dr. Holger Breyer, Dr. Hartmut Kriegs, and Dipl. phys. Rainer Schulz. They also read the manuscript and made lots of helpful comments which I gratefully accepted. I gratefully acknowledge the help of Dr. Matthias Renken in the production of many of the figures in the first part of the book. I furthermore thank Mr. Barry Rankin and my daughter Susanne Staude who helped substantially in the translation of the German manuscript. The final form of the manuscript was carefully read by Prof. Wilhelm Behrens whom I thank very much.

Finally, I would like to see that this book raises interest in the field of velocity gradient measurements and causes researchers to apply and develop measurements methods along the lines described here and to obtain new and interesting results.

Bremen, August 2001 *Wilfried Staude*

Table of Contents

Symbols XIII

1. Introduction 1

2. A Short Review of Velocity Gradient Measurement Techniques 5

2.1 The Importance of Velocity Gradient Measurement Techniques 5

2.2 The Hot Wire Method 9

2.3 Optical Methods 11

2.3.1 General Considerations 11

2.3.2 Direct Imaging 13

2.3.3 LDA: Laser Doppler Anemometry 15

2.3.4 Laser Doppler Optical Beat 17

2.3.5 The Laser Strophometry 18

2.4 Further Special Methods 19

3. Mathematical Background 23

3.1 Stochastic Quantities 23

3.2 Average Values 24

3.3 Probabilities 28

3.3.1 The Probability Density 28

3.3.2 Joint Probabilities 30

3.3.3 Conditional Probabilities 31

3.3.4 Some Properties of Probability Densities 32

3.3.5 Particles in a Vessel 34

3.4 Correlation Functions 35

3.5 The Velocity of Space and Time Dependent Stochastic Processes 40

3.6 Some Properties of Tensors 43

3.6.1 Transformation Properties of Tensors 43

3.6.2 Functions of Tensors 48

4. Physical Basis: Theoretical Part 49
4.1 Introduction 49
4.2 The Velocity Gradient 49
4.3 Motion of Particles in a Homogeneous Velocity Field 50
4.4 The Quantitative Description of Light 52
4.5 Light Scattering by Particles 55
4.6 The q-Volume of Coherence 63
4.7 The Temporal Change of the Speckle Pattern 71
4.7.1 The Motion of the Pattern 71
4.7.2 The 'Boiling' of the Speckle Pattern 77
4.7.3 'Delusive' Speckle Motion 79

5. Physical Basis: Experimental Part 83
5.1 The Illuminating Laser Beam 83
5.2 The Scattering Volume 87
5.3 The Light Scattering Particles 89
5.4 The Path of the Scattered Light 91
5.5 Stray Light 91
5.6 Test of the Set-Up 92
5.7 Measurement of Quantities Varying Randomly in Time and Space 93
5.8 From the Pattern Velocity to the Velocity Gradient 96
5.8.1 The Connection Between Pattern Velocity and $\dot{\boldsymbol{k}}_o$ 96
5.8.2 The Connection Between $\hat{\Gamma}$ and $\dot{\boldsymbol{k}}_o$ – an Example 100
5.9 Measurement of Special Components of the Velocity Gradient Tensor 103
5.9.1 Measurement of Definite Cartesian Components 103
5.9.2 Measurements of Wall Gradients 105
5.10 General Notes to Channel Flows 110

6. FRS: Forced Rayleigh Scattering 113
6.1 Basic Considerations 113
6.1.1 Introduction 113
6.1.2 The Lifetime of the Grating 115
6.1.3 Scattering Geometry and Scattering Volume 115
6.2 Experiments and Results 116
6.2.1 Experimental Set-Up 116
6.2.2 Experimental Results 117

7. LICC: Light Intensity Cross-Correlation 119
7.1 Basic Considerations 119
7.1.1 Introduction 119
7.1.2 The Demands on the Experimental Set-Up 120
7.1.3 Evaluation Using Fit Procedures 122
7.2 Experiments and Results 123

7.2.1 Experimental Set-Up 123
7.2.2 Experimental Results 124

8. DPSS: Double Pulse Speckle Strophometry 127
8.1 Basic Considerations 127
8.1.1 Introduction 127
8.1.2 The Admissible Size of the Recorded Speckle Pattern . 127
8.1.3 The Necessary Speckle Number in the Recorded Speckle Pattern 128
8.1.4 Oblique Speckle Motion and the Shape of the Scattering Volume 128
8.1.5 Special Demands on the Optical Set-Up 129
8.1.6 The Electronic Recording of the Speckle Pattern 130
8.1.7 The Determination of the Maximum of the Cross-Correlation Function 131
8.2 Experiments and Results 131
8.2.1 Experimental Set-Up 132
8.2.2 Experimental Results 132

9. DSS: Differential Speckle-Strophometry 137
9.1 Basic Considerations 137
9.1.1 Introduction 137
9.1.2 Sample Time and Separation of the Detectors 138
9.1.3 The Spatial Integration of the Light Intensity 139
9.1.4 The Measurement of the Spatial Derivative of the Speckle Pattern 139
9.1.5 The Temporal Integration of the Speckle Pattern 141
9.1.6 The Size of the Scattering Volume and the Light Power Per Coherence Area 142
9.1.7 The Shape of the Scattering Volume 144
9.1.8 Detectors and Signal Handling 144
9.1.9 Detector Noise 145
9.2 Two Evaluation Algorithms 146
9.2.1 General Requirements for the Evaluation Algorithm .. 146
9.2.2 The Weight Function $f_w(I)$ 148
9.2.3 C-Method Algorithm 149
9.2.4 S-Method Algorithm 150
9.2.5 Limits of the Method 151
9.3 Experiments and Results 152
9.3.1 Test of the Set-Up 152
9.3.2 Results of Flow Measurements 155

10. Measurement of Velocity Gradient Correlations 157
10.1 Basic Considerations 157
10.1.1 Introduction 157
10.1.2 Symmetry Considerations 159
10.1.3 Experimental Considerations 161
10.1.4 Data Evaluation 163
10.2 Experiments and Results 167

11. Outlook 173

Bibliography 175

Subject Index 177

Symbols

$\boldsymbol{x}$	vector
$\hat{X}$	tensor
$\hat{X}^T$	transpose of a tensor
$< x >$	ensemble average
$\overline{x}$	time average
x^+	dimensionless quantity
$\boldsymbol{\nabla}$	nabla operator : $\boldsymbol{\nabla} = (\frac{\partial}{\partial x}, \frac{\partial}{\partial y}, \frac{\partial}{\partial z})$
b_i	image width of lens i
c_S	normalized correlation function of stochastic process S
d_S	diameter of scattering area or scattering volume
d_C	diameter of coherence area
d_s	diameter of a volume specified by s
$\boldsymbol{e}$	unit-vector
$\boldsymbol{e}_P$	unit-vector in the direction in which the pattern velocity is measured
$\boldsymbol{e}_i$	unit-vector in the direction of the illuminating light wave
$\boldsymbol{e}_o$	unit-vector denoting the considered component of the scattered light wave
f	focal length of a lens
f_W	weight function
g_i	object width of lens i
i	imaginary unit index
j, k, l, m, n	indices
n_x	refractive index of medium x
$\boldsymbol{k}$	wave vector
$\boldsymbol{k}_i$	wave vector of illuminating light wave
$\boldsymbol{k}_o$	wave vector of a component of the scattered light wave
$\boldsymbol{q}$	scattering vector
$\boldsymbol{r}$	location vector

t	time variable
$\boldsymbol{u}$	velocity of the fluid
$\boldsymbol{v}$	velocity of the pattern of the scattered light
$w(z), w_o$	radius of a Gaussian beam
u, v, w	x-, y-, and z-components of the fluid velocity
x, y, z	orthogonal components of the location vector
A	area
C_S	correlation function of stochastic process S
D	diffusion coefficient
D_T	thermal diffusion constant
$\hat{E}$	unit matrix
$\boldsymbol{E}$	electric field
G	spatial aperture function
I	light intensity
K	turbulent energy
L	distance
N	number
$\hat{O}$	orthogonal rotation matrix
$Q(I, n)$	photon count probaility
P^1	probability density of first order
P^2	joint probability density
P_c	conditional probability density
$R(z)$	radius of curvature of a light wave
R_e	Reynolds number
S	stochastic process
T	time interval
Y	expression in the DSS technique
Z	expression in the DSS technique
α	angle, phase
β	angle, phase
	parameter describing detector sensitivity
γ_W	wall gradient
δ	distance of two detectors
η	quantum efficiency
η_S	shear viscosity
η_t	turbulent viscosity
λ	wavelength
$\boldsymbol{\kappa}$	wave vector
ν	index
ν_S	kinematic viscosity
ν_{12}	ratio of refractive indices of two media
ϕ	angle, phase
ψ	angle, phase
σ	scattering factor
$\hat{\sigma}$	stress tensor

τ	time variable
τ_c	correlation time
τ_S	sample time
τ_D	time interval
$\boldsymbol{\rho}$	location vector
ω	frequency
$\boldsymbol{\omega}$	vorticity
ω_P	angular velocity of the scattered light wave
$\hat{\Gamma}$	velocity gradient tensor

1. Introduction

Although velocity gradients play an important role in fluid dynamics, there is a lack of metrological activities in this field. One reason for this is presumably that the velocity gradient is a tensor, and therefore one has to determine generally nine independent components. Especially interesting are gradients in turbulent flows, where the gradient is heavily fluctuating. For such flows an average gradient contains very little information and only correlations of gradient components tell us something really interesting about the flow. The number of correlations, however, is much larger than the number of different gradient components.

The main reason for such little activity in this field, however, is that not very many effects are known, which can be used to measure velocity gradients. It is possible, of course, to measure the velocity at two adjacent regions in space and calculate from the two values and the known distance a velocity gradient component. However, there are no measurement techniques which allow the measurement at two 'points'. This limits such a method to determine gradients only of such structures in a flow that are large compared to the measurement volumes of the applied method.

Nevertheless, efforts have been undertaken to measure velocity gradients with a number of different techniques. Most of these will be briefly discussed in the first chapter of this book after some considerations concerning the necessity of velocity gradient measurement techniques.

The development of a method to measure all components of the velocity gradient tensor in a non-invasive way by laser light scattering by Fermigier (1980) was triggered by a publication of de Gennes in 1977. The basic idea is to write a refractive index grating in the form of a spatial plane wave into the fluid by use of an intense pulse laser. Light of a second low intensity laser will then be diffracted by this grating. If the velocity gradient in the flow does not vanish, one will generally obtain a rotation of the direction of the diffracted light. From the measured temporal change of this direction, one can calculate the size of two different components of the velocity gradient tensor in the region of the refractive index grating.

Independently of this development, students in the author's research group observed the motion of the random pattern of scattered light on a screen, when a small vessel of water containing light scattering particles was illuminated by a laser beam. This pattern motion was present as long as the convective motion within the vessel persists. The analysis of this observation showed that rather than the overall velocity of the light scattering particles a combination of different velocities is responsible for this pattern motion. This resulted in the idea to develop a method to measure velocity gradients.

On comparing the physical background of the method based on de Gennes' idea with the one based on the above observation - that is the pattern motion - it becomes obvious that it is simply the same: De Gennes' method is based on the diffraction of laser light at a regular optical inhomogeneity in form of a spatial grating, the author's on the scattering at a random optical grating. The only difference of the two methods is, that for the first one just one diffracted light wave is produced with a well defined wave vector, whereas in the second case a random but coherent superposition of such waves is produced. The temporal change of the direction of these waves, however, is exactly the same in the two cases.

This agreement on a basic physical level of the two methods is the reason why both methods are considered in this book, even though the author has no personal experience of the first one.

For the selection of topics treated in this book the author had to take into account the fact that the application of any of the presented methods is of interest mainly to the experimental researcher in the field of fluid dynamics. For the successful implementation of an experimental set-up, it is necessary to have a good understanding of the physical and mathematical background, because one cannot deem any of them a well established measurement method, which only needs to be applied. So anyone who decides to build an apparatus on these ideas can only successfully do so if he possesses a good knowledge of its mathematical and physical basis.

These considerations were the main reason for starting this book with a chapter presenting the mathematical background necessary for an understanding of the physical parts of the book. The selection of topics within this chapter is based on the author's experience with students who worked in his group. The greatest part of this chapter comprises the introduction into the basic aspects and the mathematical description of stochastic processes. Relevant terminology for the subject is also introduced here.

A small subchapter is dedicated to the tensor calculus, since a knowledge of the transformation properties of tensors in rotation is indispensable for a successful application of any one of the presented measurement methods. This part, however, assumes a certain familiarity with the concept of tensors, and does not represent an introduction to tensor calculus. To the reader not familiar with basic properties of tensors a study of an introductory book on tensor calculus is recommended.

Chapter 4: 'Physical Basis - Theoretical Part' is in a sense the central part of the book. Here, the physical effect as the basis of the measurement methods is described in detail, in particular its theoretical aspects.

In the next chapter the experimental aspects common to all measurement methods are displayed. Because of the rather complicated connection between a measured velocity component from the pattern of the scattered light, and the special component of the velocity gradient tensor responsible for this motion, an example calculation for a definite scattering geometry has been included. In the final part of this chapter the set-ups for measuring special components of the velocity gradient tensor as for instance vorticity components or wall gradients are discussed.

The following four chapters are dedicated to the detailed description of four measurement methods, which differ in the way the velocity of the moving pattern of the scattered light is measured. Some use pulsed lasers, and others continuous wave lasers. Each chapter starts with a representation of the theoretical background of the method, where special emphasis is put on its particular aspects. This part is followed by a discussion of the experimental set-up and, finally, experimental results are shown.

The last chapter is dedicated to the measurement of temporal correlation functions on the basis of one of the discussed methods. From the value of these correlation functions at time zero, the turbulent energy dissipation and the enstrophy can be calculated. The correlation functions themselves can yield information on special aspects of the turbulent flow. The last chapter is an outlook.

The author hopes that this book will make clear which quantities can be measured by the methods described, and should contain sufficient information to allow the researcher to build a set-up for the successful measurement of velocity gradients.

2. A Short Review of Velocity Gradient Measurement Techniques

2.1 The Importance of Velocity Gradient Measurement Techniques

The properties of fluid flow are to a large extent determined by the forces which exert adjacent regions onto each other. The force per area separating the two neighbouring regions is called stress $\hat{\sigma}$. Common to all fluids is an isotropic stress normal to the considered area, it is known as pressure. In ideal fluids this pressure is the only force which acts within it. Real flowing fluids exhibit also stresses within the area which is known as shear stress. In any fluid one source of this shear stress is the velocity gradient. The most simple model of a real fluid is the Newtonian fluid where the connection between velocity gradient and the stress is assumed to be linear. There is a large class of fluids where this assumption is justified. For those there are two parameters which establish the connection between velocity gradient and stress. There are, however, also fluids where the stress depends in a more complex way on the velocity gradient, especially if the fluid is composed of complex molecules or of different species of molecules. It may even explicitly depend on time. This is caused by an increase of the molecular order within the fluid caused by the velocity gradient, a possible result is for instance visco-elasticity .

Especially for more complex fluids it is important to study the connection of the velocity gradient and the produced stress. This is one of the tasks of rheology. It seems obvious that a method to measure velocity gradients will be very helpful to investigate such connections.

Another area in science where the possibility to measure velocity gradients will be very helpful is fluid mechanics. This will become obvious if we consider the theoretical description of fluid flow. The general equation of motion is the well known Navier-Stokes equation (NSE), which is a differential equation for the velocity:

$$\rho\left(\frac{\partial \boldsymbol{u}}{\partial t} + (\boldsymbol{u}\,\boldsymbol{\nabla})\,\boldsymbol{u}\right) = -\boldsymbol{\nabla} p + \boldsymbol{F}_{ext} + \boldsymbol{\nabla} : \hat{\sigma} \tag{2.1}$$

where the product $\boldsymbol{\nabla} : \hat{\sigma}$ is defined by

$$(\boldsymbol{\nabla} : \hat{\sigma})_j = \sum_i \frac{\partial \sigma_{ij}}{\partial x_j} \tag{2.2}$$

In this equation ρ is the mass density of the fluid, $\boldsymbol{F}_{ext}$ are external or body forces, for instance gravitational or electric forces, and p is the pressure.

If for a special fluid the connection between stress and velocity gradient is known the equation of motion can in principle be solved for a given set of boundary and initial conditions. However, due to the term $(\boldsymbol{u}\,\boldsymbol{\nabla})\,\boldsymbol{u}$, quadratic in the variable $\boldsymbol{u}$, solutions for a wide range of boundary conditions show chaotic behaviour, which is well known as turbulence. In those cases no analytic solutions can exist, and one has to resort to numerical calculations. However, solving the equation requires huge calculating power, especially if the flow evolution should be followed for a strech of time long enough to calculate statistical properties of the flow.

Since, however, fluid flows play an enormous role in many areas of science and industry there is a long tradition in fluid mechanics to tackle turbulence for steady flows in approximate ways which avoid the necessity to solve the Navier-Stokes equation rigorously.

To show the necessity of measuring velocity gradients let us have a somewhat closer look at some of these models. For that purpose we consider a simple incompressible Newtonian fluid without external forces. Incompressibility means

$$\boldsymbol{\nabla u} = \sum_i \frac{\partial u_i}{\partial x_i} = 0 \tag{2.3}$$

and the connection between velocity gradient and stress tensor is

$$\sigma_{ij} = \eta_S \left(\frac{\partial u_i}{\partial x_j} + \frac{\partial u_j}{\partial x_i} \right) \tag{2.4}$$

where η_S is the shear viscosity. We then obtain the following special form of the NSE, written down for an arbitrary component i

$$\frac{\partial u_i}{\partial t} + u_j \frac{\partial u_i}{\partial x_j} = -\frac{1}{\rho} \cdot \frac{\partial p}{\partial x_i} + \nu_S \nabla^2 u_i \tag{2.5}$$

where we tacitly assumed the summation convention which means that in any term where an index appears twice this index has to be summed over. ν_S is the so called kinematic viscosity, defined by

$$\nu_S = \frac{\eta_S}{\rho} \tag{2.6}$$

For a theoretical description of stationary flows usually a decomposition (Reynolds decomposition) of the variables $\boldsymbol{v}$ and p is performed in time averaged and fluctuating parts, respectively:

$$\begin{aligned} \boldsymbol{u} &= \overline{\boldsymbol{u}} + \widetilde{\boldsymbol{u}} \\ p &= \overline{p} + \widetilde{p} \end{aligned} \tag{2.7}$$

Inserting these quantities in the NSE and averaging leads to

$$\overline{u}_j \frac{\partial \overline{u}_i}{\partial x_j} + \frac{\partial}{\partial x_j} \overline{u_i u_j} = -\frac{1}{\rho} \cdot \frac{\partial \overline{p}}{\partial x_i} + \nu_S \nabla^2 \overline{u}_i \tag{2.8}$$

If this equation is subtracted from equation (2.5) one obtains an equation for the fluctuating variables:

$$\frac{\partial \widetilde{u}_i}{\partial t} + \frac{\partial \overline{u}_i}{\partial x_j} \widetilde{u}_j + \frac{\partial \overline{u}_j}{\partial x_j} \widetilde{u}_i + \frac{\partial}{x_j} \left(\widetilde{u}_i \widetilde{u}_j - \overline{u_i u_j} \right) = -\frac{1}{\rho} \frac{\partial \widetilde{p}}{\partial x_i} + \nu_S \nabla^2 \widetilde{u}_i \tag{2.9}$$

Of course, the time average of any of the terms in this equation is zero.

We have now obtained a system of the two equations (2.8) and (2.9), but this system is not very helpful since in addition to the two variables $\overline{\boldsymbol{u}}$ and $\widetilde{u}_i$ the new variable $\overline{u_j u_l}$ appears, the so called Reynolds stress. A new equation can be obtained if the last equation is multiplied by $\widetilde{u}_m$ and the time averaging is carried out. But again, new variables appear, for instance $\overline{u_j u_l u_m}$. So one is always left with more variables than equations. This is known as the closure problem in the theory of turbulence.

In order to obtain a system of equations which can be solved unambiguously one has to find additional equations in order to 'close' the system of equations. Because the required additional equations cannot be derived from the basic NSE itself one has to find them in an intuitive way.

One of the most simple approaches is to assume a proportionality between the Reynolds stress and the mean velocity gradient:

$$\sigma_{ij} = \rho \overline{v_i u_j} = \rho \eta_t \frac{\partial \overline{u}_i}{\partial x_j} \tag{2.10}$$

where η_T is a parameter, the so called 'eddy viscosity', which has to be determined empirically.

Another famous approach represents the mixing length model, where the important parameter is the mixing length l_m. This model has proved valuable especially for jets and shear flows. Here the dependence of η_t on l_m and a typical mean-velocity difference v_δ across the layer or the jet according to

$$\eta_t = l_m \cdot u_\delta \tag{2.11}$$

is assumed. The mixing length l_m is assumed to scale in the two mentioned types of flow with the layer thickness or the distance from the wall, respectively.

A closure procedure widely applied to engineering problems is the so-called K-ϵ-model, which is based on a connection between the turbulent energy K and the dissipation rate of the turbulent energy ϵ:

$$K = \rho \overline{u_i u_i} \tag{2.12}$$

$$\epsilon = \frac{\nu_S}{2} \overline{\left(\frac{\partial u_i}{\partial x_j} + \frac{\partial u_j}{\partial x_i} \right)^2}$$

This connection is expressed by the equation

$$\eta_t = \rho C_T \frac{K^2}{\epsilon} \tag{2.13}$$

where C_T is a constant. In this approach two new equations must be found for the two new variables K and ϵ. They have the same structure as the NSE. Although both equations are coupled they are scalar equations and can be solved much easier than the NSE's.

Of course, all these models are based on assumptions the validity of which has to be checked. They usually work well for special types of flows but more or less fail for others.

It would therefore be extremely valuable if different terms in the equations could actually be measured for different flow types. That would not only allow to test the validity of the basic assumptions but also to open the chance for the development of improved models.

The fact that most of these terms consist of the variables velocity and velocity gradient and correlations of them - i.e. average values of products - underlines the necessity to measure just these quantities. Whereas reliable and non-invasive techniques to measure velocities are widely used - think of the famous Laser-Doppler technique (c.f. Sect. 2.3.3)- the methods described in this book to measure velocity gradients and their correlations by two research groups is in our opinion able to fill this gap.

For the description of turbulent flows a quantity plays an important role which is directly connected with the velocity gradient: the vorticity $\boldsymbol{\omega}$. It describes quantitatively the rotational motion, which is characteristic of turbulence. Mathematically the vorticity is the 'rotation of the velocity'

$$\boldsymbol{\omega} = \boldsymbol{\nabla} \times \boldsymbol{u} \tag{2.14}$$

Apparently, the components of the vorticity are linear combinations of components of the velocity gradient. One obtains the vorticity equation by taking the rotation of each term of equation (2.8). For incompressible fluids one obtains after some mathematical manipulations:

$$\overline{u}_l \frac{\partial \overline{\omega}_j}{\partial x_l} = -\overline{\widetilde{u}_l \frac{\partial \widetilde{\omega}_j}{\partial x_l}} + \overline{\widetilde{\omega}_l \frac{\partial \widetilde{u}_j}{\partial x_l}} + \overline{\omega}_l \frac{\partial \widetilde{u}_j}{\partial x_l} + \nu_S \frac{\partial^2 \overline{\omega}_j}{\partial x_l \partial x_l} \tag{2.15}$$

The remarkable aspect of this equation is the fact that no pressure term occurs, since the rotation of the pressure term in the NSE is identically zero. Each term in this important equation contains a velocity gradient component. Therefore, for the experimental verification of theoretical models based on this equation again velocity gradient data are required.

A further important aspect of the phenomenon turbulence is the so called 'intermittency'. Experimental observations of turbulent flows reveal that on increasing the velocity of the fluid in any experiment the turbulent behaviour starts with stretches of irregular motion interrupted by those of rather smooth, laminar motion. With increasing flow velocity the laminar periods become shorter and larger portions of the fluid are 'really' turbulent. But even at highest velocities there appear regions with regular mainly rotational motion and a certain life-time, which have special shapes and are called 'coherent structures' . This property of turbulent flows cannot be simply inferred from an inspection of the NSE or the equations derived thereof. It seems obvious that this intermittency has a strong influence on the properties of a flow although it is difficult to quantify it both, experimentally and theoretically. Theoretical investigations show that the intermittency has a strong influence on the probability distribution of velocity gradient values. It is therefore not only of interest to measure velocity gradient correlations but also distribution functions.

An important role in the theory of turbulence play spatio-temporal correlation functions of the velocity, which are better known as two-point correlation functions. Little attention has been paid so far, however, to the study of respective correlation functions of the velocity gradient. One can easily imagine that statistical properties of coherent structures will show up in these functions much more directly.

A special velocity gradient which for the engineer doing applied research is of high importance is the so called wall gradient γ_W, the quantity which describes the increase of the flow velocity normal to and directly at a solid boundary. This γ_W is responsible for the friction forces which exert the walls on the fluid and tend to bring it to rest. Of course, not only the average value of this quantity is of interest but also the size of its fluctuation. The measurement of the wall gradient, however, is especially difficult. In most flow situations one has to measure extremely close to the wall in order to really obtain this quantity reliably. Furthermore, the measurement technique used must by any means be a non-invasive one if the flow at the wall should in no way be disturbed. Here again, the methods described in this book allow measurements of this quantity.

2.2 The Hot Wire Method

The oldest method to measure the velocity in a fluid with high temporal and spatial resolution is without any doubt the hot wire technique. Until these days it plays an important role in experimental fluid mechanics. The basic physical effect it is based on is the velocity dependent heat transport away from an electrically heated wire. If the current through the wire is kept constant the temperature of the wire is a function of the fluid velocity provided the temperature of the fluid is constant. Usually a tungsten wire with temperature dependent resistance is used. It is possible to use the wire as one resistor in a Wheatstone bridge, which allows the measurement of very small resistance changes. Often, however, an electronic circuit is used which controls the current through the wire in a way that the resistance of the wire and with it its temperature is kept constant. The size of the current then is a measure of the fluid velocity. The length and diameter of the wire can be made as small as 1 mm and 50μm, respectively. The wire is generally fixed to the ends of a metal prong. Since the relation between heat transport and fluid velocity is dependent on many parameters one generally has to gauge any probe. If the ratio length/diameter is larger about 250 one can neglect the cooling effect for the velocity component parallel to the wire. Nevertheless, one simple wire cannot deliver information about the direction of the velocity within the plane perpendicular to the wire axis. To measure all components of the flow velocity one therefore needs probes with several wires.

Already half a century ago measurements of velocity gradient components using probes with several wires were performed. The idea in these experiments is to use clustered arrangements of wires, where the different clusters are used to measure velocities at different locations in space. The sensors have progressively been improved, and up to 9 wires were incorporated. Figure 2.1 shows an arrangement of clusters used by P. Vukoslavčević et al. ([1]). The wires are inclined to the plane shown, the mid points of the wires lie in the same plane. With this arrangement velocity differences in that plane can be calculated from the signals of the individual wires and from these six different gradient components.

Those velocity gradient components which describe the change of the velocity in the main flow direction - perpendicular to the plane shown - are calculated according to the Taylor hypothesis from the temporal change of the velocity. According to this hypothesis the spatial derivative of the velocity component u_x in the main flow direction x, i.e. $\frac{du_i}{dx}$ can be replaced by $\frac{1}{u_x}\frac{du_i}{dt}$.

Very extensive measurements of gradients were done by Balint et al, [1] who not only measured all mean velocity gradient components in a turbulent air channel flow but also variances, correlations, and correlation functions of these quantities.

The most outstanding advantage of the hot wire method is no doubt the possibility to measure the temporal development of the velocity with high resolution and calculate from these data other interesting properties as for instance correlation functions but also probability distributions and joint probability distributions of velocities. Of course, with the mentioned sensors it is also possible to measure analogous quantities of velocity gradients. A further advantage over optical methods is the fact, that no additional light scattering particles must be added to the fluid (see below). So especially for the application to gas flows one directly measures the velocity of the gas and not that one of the particles within the gas.

There must, however, be envisaged two severe disadvantages of this method. On the one hand, each probe has to be gauged with a rather extensive procedure in order to get reliable velocity results. As an illustration, the mentioned work of Balint et al. consists of two parts with almost the same

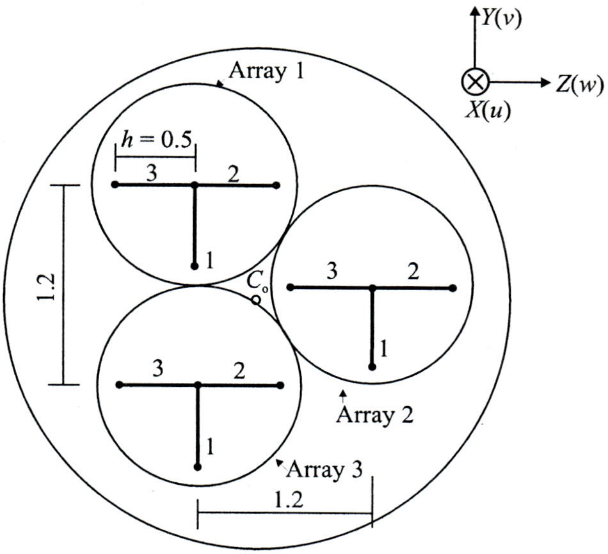

Fig. 2.1. Schematic top view of the hot wire probe developed by P. Vukoslavčević et al.[1].The sensor consists of three arrays of three wires each. The three wires of each array have one contact in common, the end of the wires lie below the x-y plane, they form an angle of 45° with that plane. Figure taken from [1]

number of pages. The first part is exclusively devoted to the construction and especially to the gauging procedure and test measurements.

On the other hand, more severe is the fact that the insertion of the probe in the flow more or less disturbs it, so that actually velocities of the changed flow are measured. Generally, the influences are less severe in gas flows than in liquids, therefore this method is more widely used for the studies in wind channels. Especially doubtful is the application of this method in flows without a predominant flow direction, i.e. a flow where locally the flow direction changes considerably. It is obvious that in cases where the prongs point upflow the flow pattern is extremely disturbed. A special case of such a situation is near a stagnation point, where the flow is very sensitive to disturbances.

2.3 Optical Methods

2.3.1 General Considerations

All optical methods are based on the investigation of light scattered off the flowing fluid. Common to all techniques is the requirement of optical inhomogeneities within the fluid. Even pure fluids - liquids as well as gases - have a certain light scattering power because of their corpuscular structure. However, the fraction of light scattered in volumes of a cubic milli-meter, say, is extremely small so that a measurement method based on the evaluation of light scattered in small volumes of a pure fluid is not feasible. Therefore, light scattering particles are added to the flow, which procedure is called seeding. The requirement that the particles follow the flow with high fidelity is best fulfilled if the mass densities of the particles and the fluid are identical. Only for particles large compared to the scale on which the velocity of the flow changes there will result additional forces which may cause particle paths different from stream lines. It is obvious from basic mechanics that this slip between the fluid and the particles increases with increasing acceleration. In gases the equality of mass densities cannot be achieved. In that case the slip can only be reduced by using smaller particles.

For measurement techniques which apply no direct imaging of the region of interest one has to make sure that the scattered light which is to be analysed originates in a certain small volume around the point of interest, the so called scattering - or measurement - volume. This volume is determined on the one hand by the spatial extension of the illuminating light wave and on the other hand of the light collecting optics on the detection side. In a more rigorous way one has to speak of a spatial aperture function which gives the relative amplitude of the light wave at the detector which results from a particle at a definite location. Nevertheless, for practical purposes the notion of the scattering volume is widely used and quite helpful. As a rule, the spatial resolution of such a method is restricted to the size of that volume.

Some methods require spherical seeding particles such that the scattered light wave is only affected by the position of the particles and not by their orientation in space.

Generally, velocity information on the scatterer can only be contained in the temporal change of the scattered light wave. For that reason any method to measure velocities or quantities derived from it - as velocity gradients - must analyse the scattered light at least at two instants of time or during a certain period.

Considering the task of measuring velocity gradients it is obvious that the light wave scattered by one single spherical particle cannot carry enough information, namely the velocity difference at different locations in space. This requires at least two particles which during the measurement process should lie within a region where the velocity gradient can be assumed constant. We will henceforth call this volume 'homogeneity volume' , because a flow with constant velocity gradient is called homogeneous. This consideration makes clear that for practical purposes a minimum concentration of seeding particles is required such that within the scattering volume - which should not be larger than the homogeneity volume - at least two particles are present at almost any time.

The different optical measurement techniques can be divided into those which are based on methods analysing light scattered by single particles (e.g. Laser-Doppler-Anemometry LDA, direct imaging methods) and those which evaluate light scattered by ensembles of particles (e.g. scattered light correlation techniques).

The methods mentioned first generally have an optimum seeding concentration independent of particle size, although it will depend on the size of the scattering volume, the maximum velocity, and other parameters. Furthermore, the light scattering power must not be too small in order to analyse the light scattered by any single particle.

The requirements for the second type of methods are less specific. It is just required that a minimum amount of light is scattered within the scattering volume. Consequently, one can use here very small particles provided their concentration is large enough. These methods are then especially well suited for investigations of flows where for larger particles the risk of slip cannot be excluded.

A second way to group the measurement methods is according to whether they are direct or indirect methods. One would call a method direct if the measured quantity directly depends on the quantity of interest, i.e. in our case on the velocity gradient. Indirect methods have to measure several quantities in order to get the desired result. So for instance most of the optical methods presented so far to measure velocity gradients are indirect methods: They are used to measure either the velocity at different regions in space or the velocity difference. From those data and the distance of the two regions the gradient can then be calculated. It seems obvious that direct methods have the advantage that different data with certain errors need not be combined, which according to the laws of error propagation leads to increased errors in the final result.

2.3.2 Direct Imaging

With these methods one obtains velocity information from large regions, generally two-dimensional ones, but also more complex methods have been developed which allow the holographic recording of the velocity of a three-dimensional region.

A typical set-up of the widespread light-sheet technique is shown in Fig. 2.2. By use of cylindrical lenses a light beam is formed with a very high ratio of height to width, the light scattered by the particles within this sheet is then recorded. Several techniques are used to obtain velocity information.

The most widely applied technique is to use a double pulse laser, illuminating with it the interesting region in the flow and taking one photograph. Each particle is then represented on the picture by two spots. From the picture one can then obtain the direction and absolute value of the velocity in that direction at many points in the flow. To determine the sign of the velocity one can for instance move the second image with respect to the first one

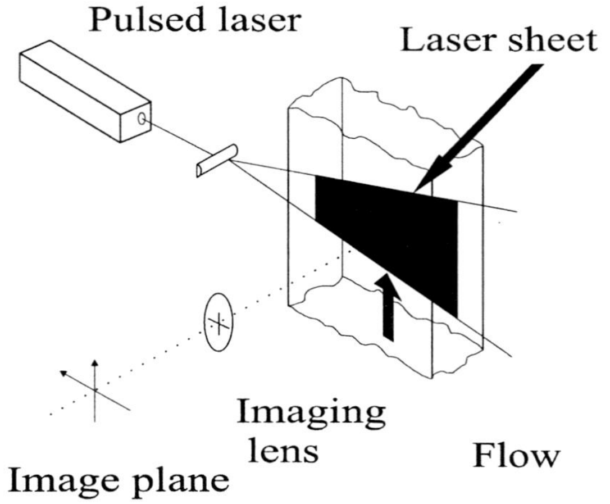

Fig. 2.2. Typical optical set-up for two-dimensional direct imaging. Figure taken from [2]

by use of a rotating mirror. If the distance of the image spots of any particle is larger than their distance caused by the flow velocity the determination of the velocity vector within the plane of the light sheet becomes unambiguous.

With the ever growing increase in the resolving power of electronic cameras photographic recording has to a certain extent been replaced by electronic recording, which also favours computer based data evaluation. But photographic recording with its larger resolution is still used, especially with holographic methods.

The seeding of the flow can be optimized: With a small particle concentration the spatial resolution of the method is low and calculated velocity gradient values are not reliable. But with concentrations too high the interpretation of the images becomes ambiguous because of the overlap of the spots of different particles.

A large amount of publications shows that from properly performed experiments velocity maps, and from those vorticity maps can be obtained. Figure 2.3 shows an example of a measured vorticity map in a turbulent pipe flow. Average values for stationary flows have been obtained by taking many recordings (e. g. 60 in [3]). It must be mentioned, however, that from 60 velocity or velocity gradient values no reliable average values in turbulent flows can be calculated. The mentioned measurements were done on a channel flow, and in addition to the average over the pictures a spatial averaging over locations with the same wall distance were carried out.

The direct imaging methods have the big advantage over 'point' methods as LDA that they deliver spatial information on the flow. So they are very helpful to study for instance the general shapes of flows and shapes of coherent structures. The investigation of the temporal behaviour, however, would

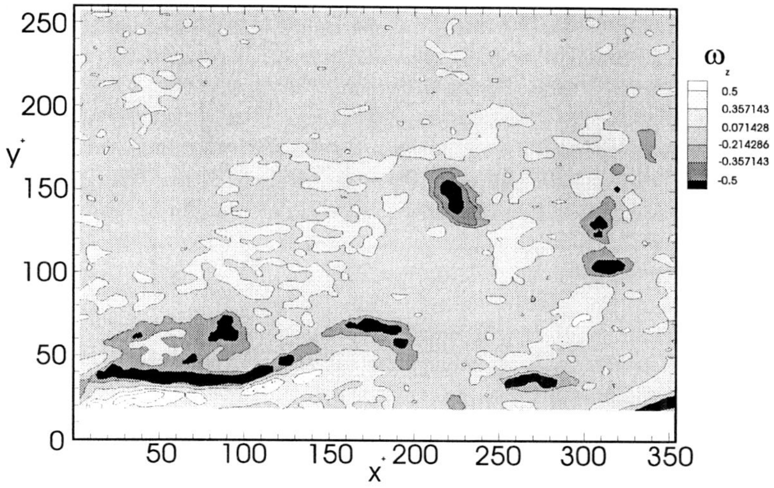

Fig. 2.3. Map of the instantaneous z-component ω_z^+ of the vorticity in a turbulent channel flow. The channel wall is at $y^+ = 0$, the flow direction is along the x-axis. The index $^+$ denotes that the quantities are scaled (c.f. Sect. 5.10). Figure taken from [3].

require an enormous - mainly computational - effort. Moreover, the time necessary to transfer the information of a high resolution camera to a computer is still not sufficiently high. Nevertheless, these methods will definitely become very important in the future.

2.3.3 LDA: Laser Doppler Anemometry

In the last few decades the Laser-Doppler Anemometry has become an extremely powerful and widely used measurement technique in fluid flow research. It is based on the optical Doppler effect: If a moving light scattering object is illuminated by a monochromatic light wave, the scattered light suffers a frequency shift (Doppler effect) depending on the scattering geometry and the size of one velocity component of the object. If the moving object is illuminated with two coherent light waves traveling in different directions the scattered wave generally shows a periodic amplitude modulation. The frequency of this modulation is directly proportional to one component of the velocity. This effect can be explained in two different ways:

- The two plane light waves interfere in their overlap region and form a sinusoidal intensity pattern as shown schematically in Fig. 2.4. A particle within that region scatters light proportionally to the local intensity from which immediately follows that a moving particle with a velocity component normal to the intensity planes causes a sinusoidally varying scattered light intensity.

- For fixed observation direction and particle velocity, the Doppler shift of the scattered light depends on the direction of the illuminating light wave. A theoretical treatment shows that the difference of the Doppler shifts of the two scattered waves is independent of the observation direction. Due to their interference the detector signal is modulated with that beat frequency. Figure 2.5 shows an idealized detector signal.

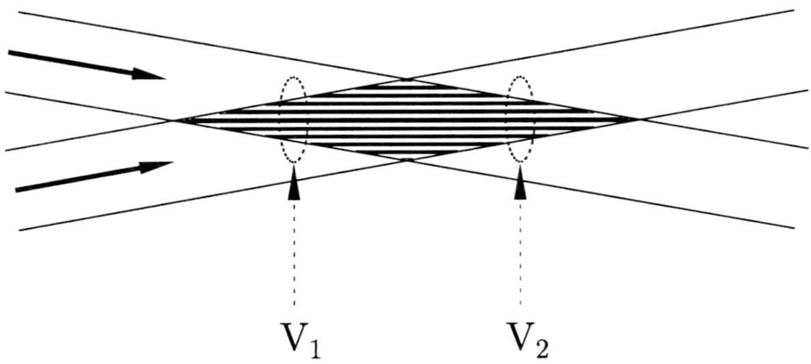

Fig. 2.4. The light intensity in the region of the crossing of two coherent plane waves. Two scattering volumes are indicated which allow the determination of one velocity gradient component

The detector signal does not contain information on the sign of the velocity, i.e. two particle moving exactly opposite to each other produce the same signals. A method often used to obtain the motion direction is to shift the frequency of one of the illuminating light waves. The intensity pattern in the overlap region is then traveling in one direction. A particle at rest now produces a sinusoidal detector signal. The frequency of the signal is raised or lowered depending on the motion direction. This technique delivers unambiguous results if the highest occurring particle velocity is still lower than the velocity of the moving intensity pattern. Velocity gradients are determined with LDA in the following way: Two different scattering volumes within the overlap region are viewed. In a set-up chosen by J.H. Agüi and J. Andreopoulos [4] the two scattering volumes are chosen as indicated in Fig. 2.4. If two particles traverse these two volumes at almost the same time one obtains two velocity values from which a definite component of the velocity gradient can be calculated.

Unfortunately, different requirements for an 'ideal' set-up tend to push certain parameters in opposite directions: Generally, the distance of the two scattering volumes should not be larger than the homogeneity volume. In order to obtain an accurate velocity gradient value one has to know the distance of the two particles fairly accurately. This requires scattering volumes small compared to their distance. However, accurate velocity values can only

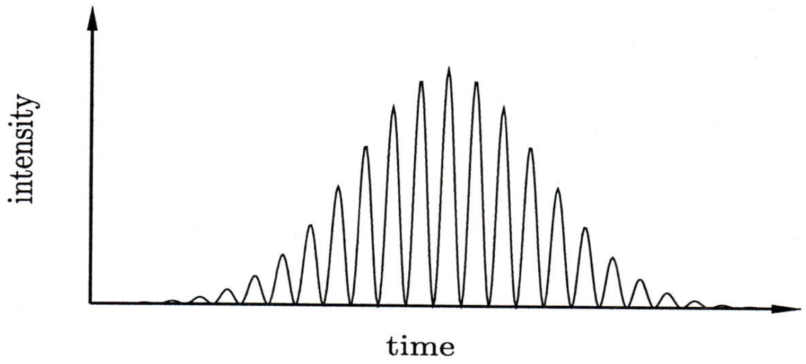

Fig. 2.5. The scattered intensity produced by a particle traveling through the fringe pattern shown in Fig. 2.4

be obtained if a particle within the scattering volume traverses enough interference fringes. That would require a scattering volume rather large. For the set-up shown in the figure only such particle velocities can be determined reliably which have much larger components along the the long axis of the scattering volumes than perpendicular to it. To obtain reliable velocity values of particles traveling almost parallel to the short axis, i.e. almost parallel to the intensity planes, would apparently require a much larger extension of the scattering volumes in that direction than shown in the figure. But then the distance of two particles in the two scattering volumes could not be determined reliably. This consideration is especially important if the direction of the velocity fluctuates appreciably. So the method is likely to work well for flows with a predominant velocity component in a fixed direction. Since LDA is basically a 'one particle measurement scheme' the seeding of the fluid must be adjusted carefully to the size of the scattering volume.

2.3.4 Laser Doppler Optical Beat

As mentioned in the last subsection the velocity of particles simply leads to changes of the frequency of the scattered light waves. If therefore the light scattered by two particles with different velocities is superimposed an oscillating intensity results with a beat frequency which varies linearly with the velocity difference. Asakura [5] was the first who used this phenomenon to measure velocity gradients of a rotating ground disk. Later Ötügen et. al. [6] applied this method to measure the velocity gradient within a rotating fluid, results of their measurements are shown in Fig. 2.6. From a physical point of view the difference between this method and the one based on LDA is the following: With LDA values of the velocity at two locations in space are determined separately and they are combined numerically to get the desired gradient value. With laser doppler optical beat the scattered light itself 'carries out the calculation' of the velocity difference. The requirements for the separation of the two scattering volumes and their size is the same as with the LDA method.

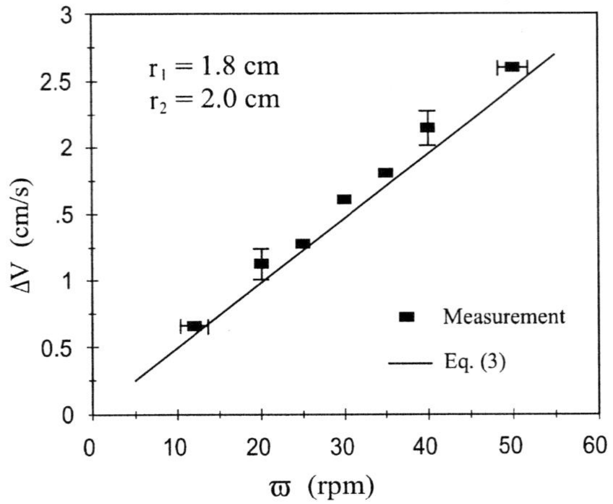

Fig. 2.6. Measured velocity difference ΔV in a laminar rotational flow, theoretical and experimental data. ω is the rotation frequency of the circular tube, r_1 and r_2 denote the distance of the two scattering volumes from the rotation axis. Figure taken from [6].

2.3.5 The Laser Strophometry

Consider the light wave produced by scattering of a coherent plane wave within a certain volume. This wave contains all the information on the optical inhomogeneity of this volume down to dimensions of the order of the wavelength. However, about half of the information is encoded in the phase of the scattered wave and can only be retrieved by sophisticated methods as for instance holography. In many cases one resigns oneself to the information contained in the phase. But there is still liberty to modify the scattered wave in order to obtain special information on the scatterer. The most common way is direct imaging which essentially leads to a loss of the information on the depth within the considered volume. If one, as an alternative, takes a picture in the focal plane of the imaging system one obtains the Fourier transform of the scattered wave. It is not surprising that in the amplitude of this image other information on the scatterer is encoded than in the direct image. In turns out that for instance a translation of the object in an arbitrary direction does not change this image, a rotation or distortion, however, leads to notable changes.

This consideration elucidates the advantage of the last mentioned type of optical manipulation of the scattered light wave for velocity gradient measurements. The velocity of the object is transformed entirely into the phase of the image wave in the focal plane and is therefore not observed.

Unfortunately, there is further information encoded in the phase which is desirable for velocity gradient measurement methods. This is not surprising. The velocity gradient describes rotation and distortion in three dimensions whereas the image is two-dimensional.

The theoretical treatment, which is explained in this book shows that the lateral motion of the Fourier image allows to determine two independent components of the velocity gradient tensor in one light scattering experiment, i.e. in an experiment which uses one plane illuminating light wave and analyses the light scattered into a small spherical angle. The two gradient components simply determine the lateral shift of the Fourier image.

A discussion of the advantages and disadvantages of measurement methods based on this effect is postponed to the final chapter of this book. It should be mentioned here as a final remark that all these methods are direct ones, i. e. only **one** quantity has to be measured in order to get **one** velocity gradient value.

2.4 Further Special Methods

We will now briefly discuss further methods which have been used to measure special velocity gradients.

The first one was presented by M.B.Frish and W.W. Webb [7]. and is an optical method to measure the vorticity. The idea is to use as seeding particles small bodies with plane reflecting faces. If such particles are illuminated by a plane light wave the scattered light shows high intensities in directions where specular reflection at those faces occurs. A vorticity within the flow will rotate the particles and, consequently, also those specular reflected rays will perform a rotation. On a screen one will then observe a motion of bright spots. The authors measured the motion of these reflexes by placing two parallel slits with a detector behind each of them in the scattered light field. The signals in the two detection channels are shifted just by the time needed for a reflected ray to travel from one slit to the other. Figure 2.7 shows vorticity profiles in a Poiseuille flow measured with this technique,

A special optical method to measure wall gradients has been developed by A.A. Naqwi and W.C. Reynolds [8]. Their method is similar to LDA but the interference pattern they use is essentially the one shown in Fig. 2.8 (denoted 'Fan Fringes'), the pattern being the same in the direction normal to the drawing plane. Sufficiently close to a wall the flow velocity is parallel to it and it depends linearly on the distance from the wall. Consequently, the gradient in that region is constant, it is called wall gradient. It is then obvious that any particle traveling through that special light field produces scattered light with an intensity oscillating with the same frequency: The frequency is proportional to the particle velocity v and inversely proportional to the

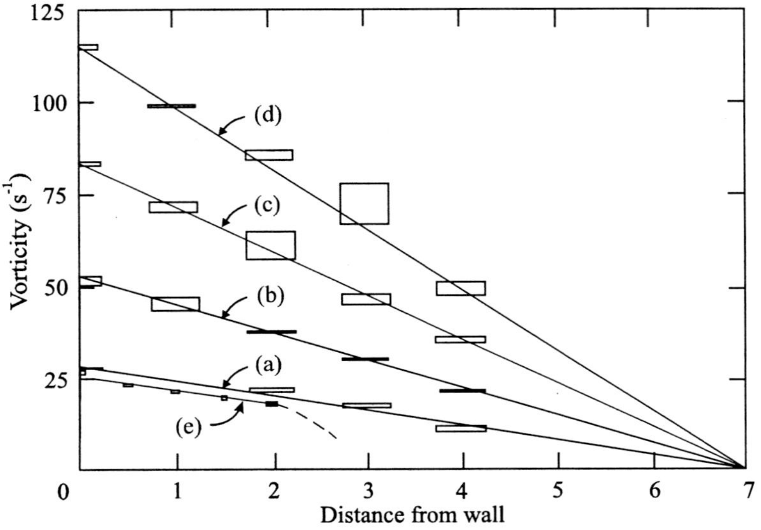

Fig. 2.7. Measured vorticity profiles in Poiseuilles flow. Calculated speeds: (a) 2.9 cm s^{-1};(b) 4.8 cm s^{-1}; (c) 8.6 cm s^{-1};(d) 13.0 cm s^{-1};(e) 10.3 cm s^{-1} (not fully developed Poiseuille flow). Figure taken from [7].

spacing Λ of the fringes. If we denote the distance from the wall by δ we have $v \sim \delta$ and $\Lambda \sim \delta$ from which immediately follows that $\omega \sim v/\Lambda$ is independent of δ.

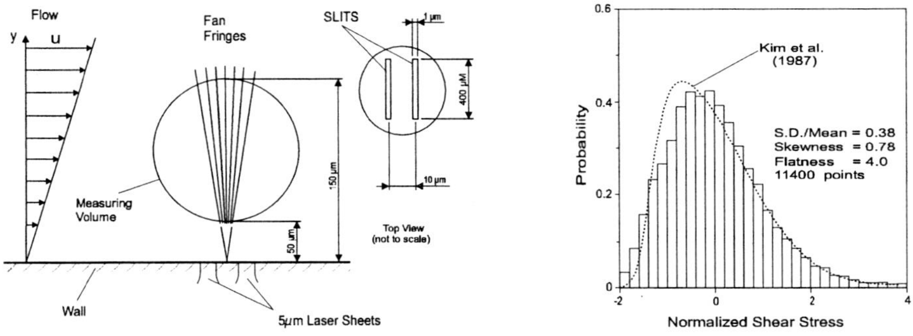

Fig. 2.8. Left: Flow velocity, interference pattern with measuring (scattering) volume, and slit dimensions. Right: Measured distribution of wall gradient. Figure taken from [8].

The required intensity pattern can only approximately be obtained. The authors achieved that in the following way: Two cylindrical light waves are produced by illuminating two parallel narrow slits (width $\approx 1\mu$m, 10 μm apart, see Fig. 2.8) with coherent light in the plane of the flow boundary.

The resulting interference pattern is hyperbolic, but at a distance larger than about 50 μm from the wall it has to a good approximation the desired shape.

One further method should be mentioned which is, however, only applicable to laminar flows with known flow geometry. It is based on the fact that the correlation function of the light scattered in a fluid drops off much faster and also has another time dependence if within the fluid a velocity gradient exists. This method was presented by Fuller et al. [9].

3. Mathematical Background

3.1 Stochastic Quantities

A theoretical treatment of the methods considered in this book reveals that random fluctuating quantities play a major role in each of them. On the lowest level, the microscopic one, we have the diffusion of the molecules of the fluid. Even in a laminar flow these elementary particles of the fluid do not move along the stream lines which only characterize the averaged velocity of these particles. Consequently, these streamlines only describe the macroscopic velocity field.

Three of the four measurement methods presented here require the addition of light scattering particles to the fluid. These particles are very much bigger than the fluid molecules, and the 'randomness' of their motion, which is known as Brownian motion, is less pronounced than the one of the molecules. That means they follow the stream lines much better than the molecules, provided their mass density is more or less the same as that of the fluid. Nevertheless, the stochastic behaviour of the particles has to be taken into account if one wants to describe satisfactorily the properties of the scattered light. Since all methods are based on the interference of the light scattered off these particles, it is obvious that a deviation of their path from a stream line in the order of a wavelength ($\approx$ 0,5 μm), can lead to drastic changes of the scattered light wave.

A further source of random behaviour in connection with these methods results from the random position of light scattering particles within the fluid, that is the volume from which the scattered light results. This randomness leads to a spatial random pattern on a screen, on which this pattern can be viewed. Furthermore, the random flux of particles into and out of the scattering volume leads to a random temporal change of the pattern of the scattered light.

The above mentioned sources of randomness represent obstacles with regard to the aim of the measurement methods, i.e. they render the measurement of the quantity of interest more difficult. A major aspect of the methods considered here is the aim to measure this quantity reliably in as short a time period as possible.

One source of randomness, however, is central to the application of these methods and represents therefore in no way a disturbance. This is the random fluctuation of the velocity field in case of a turbulent flow. In this case one wants to know as much as possible about the stochastic properties of this velocity field.

A further source of randomness, which we have to consider here and which stems from quite a different origin, results from the light detection process. For some of the methods, the intensity of the scattered light is measured by counting photons. These particles represent the smallest amounts (quanta) of energy which can by detected by a light measuring process. In cases of very big light intensities one can disregard the quantum nature of light: The number of detected photons is so large that the detection system does not discriminate between one detected photon more or less. The methods described in Chaps. 7, 9 and 10, however, are based on the measurement of the light energy in very small time intervals, in which the number of detected photons is very small (experiments were done with average photon numbers down to about 5). In this case, one has to take into account that the light detection process itself is a random process. That means that even in the case where an extremely constant light wave of a stabilized laser hits the detector, one observes an entirely random series of output pulses. For known light intensity and detector efficiency one can only state the average number of detected photons in a given time interval. The noise of the detector signal produced by this effect is known as shot noise.

All mentioned randomly fluctuating quantities - particle positions, intensity of the scattered light, velocity field in turbulent flow, photon numbers - represent stochastic processes, i.e. they are time dependent quantities, but their temporal evolution is not given in a deterministic way. Some of them are furthermore space dependent - the intensity field of the scattered light and the velocity field - and this dependence in these two cases are also stochastic.

3.2 Average Values

If stochastic quantities are measured using a single measurement process, which delivers a definite value, it is apparently of little particular use, especially if this quantity shows a large amount of fluctuation. To obtain more information on such a quantity requires repetitive measurements. With a series of the obtained measurement values one can then calculate different quantities. One is often interested in the mean value of the measured quantity, but also in the variance or so called higher moments. All these quantities have in common that they are average values, or, expressed more exactly, they are estimations of average values. Exact average values can only be obtained for an infinite series of measurement values. There exist different methods to obtain average values, which are also of different relevance for experimentalists and theoreticians.

The most important ways to obtain average values shall now be made clear for the case of particles in a fluid, which is contained in a vessel. One question of interest in connection with the methods described here is the following: How far will a particle within the fluid move in a certain time τ? If we denote the location of the particle at time t by $\boldsymbol{r}(t)$ then we can express this question in the mathematical form: How large is the size of the vector $\boldsymbol{r}(t+\tau)-\boldsymbol{r}(t)$? A way to avoid the square root expression in calculating the size of a vector is to look at the so called mean square displacement of the particle:

$$S^2(t;\tau) = |\boldsymbol{r}(t) - \boldsymbol{r}(t+\tau)|^2 \tag{3.1}$$

It should be kept in mind that in this context the symbol $\boldsymbol{r}(t)$ denotes the position of the particle at a fixed time t. It is therefore advisable not to connect with the symbol $\boldsymbol{r}(t)$ the meaning: 'position of the particle as a function of time' but rather 'Position of the particle at a fixed time t'. With that meaning the t in $\boldsymbol{r}(t)$ is to be understood as an index.

The other stochastic quantity, which is of major interest for our considerations, is the intensity of the scattered light. We will describe it as a two dimensional pattern $I(\boldsymbol{r},t)$ on an observation screen. In this chapter we consider the simple average value of $I(\boldsymbol{r},t)$, whereas in later chapters, expressions which contain two or three intensity values at different times and locations will play an important role.

We will now look for an answer to the question: How can the average values of the mentioned quantities be calculated? There exist different possibilities.

Temporal Average. Let us consider first the above mentioned intensity pattern of the scattered light. In order to obtain the mean intensity at a particular point on the screen, the temporal evolution $I(t)$ of the intensity at this point is measured for a sufficiently long period T, starting at an arbitrary time t_o. Then, the following expression is calculated:

$$\overline{I}^T = \frac{1}{T}\int_{t_o}^{t_o+T} I(t)dt \tag{3.2}$$

This integral represents the so called temporal average. It is the average, which in most cases is formed when averages are determined experimentally.

Normally, the value of this integral will become essentially independent of T with increasing T. For that reason, one will generally leave off the upper index T at the quantity $\overline{I}^T$. One then uses the symbol $\overline{I}$ for a time average, even though in a rigorous sense this average has only a definite value in the limit $T \to \infty$. We will here assume the more experimentally motivated point of view and leave off the upper index, although on many occasions the experimentally determined 'average value' is by no means a fixed quantity, but also a stochastic one. It does, however, fluctuate much less than the basic quantity.

However, the value of the integral in equation (3.2) can depend on the starting time t_o, e.g. in the case of a changing laser power, or if the light scattering particles in the liquid slowly precipitate to the bottom, thus reducing the intensity of the scattered light. Where the integral is independent of the starting time t_o, one calls the stochastic process $I(\boldsymbol{r}, t)$ 'stationary'. (A more rigorous definition of the term 'stationarity' is given in Sect. 3.3.4.

We will now consider only stationary processes or, more accurately expressed, such processes which we assume to be stationary.

Let us consider, as a second example, the determination of the mean square displacement. For that purpose, one could follow the path $\boldsymbol{r}(t)$ of a single particle for a long time. The time averaged square displacement is then given by

$$\overline{S^2(t_o; \tau)}^T = \frac{1}{T} \cdot \int_{t_o}^{t_o+T} |\boldsymbol{r}(t) - \boldsymbol{r}(t+\tau)|^2 dt \tag{3.3}$$

$$= \frac{1}{T} \cdot \int_{t_o}^{t_o+T} S^2(t; \tau) dt$$

In this example, one again has to choose the averaging time T long enough, so that the value of the integral is to a good approximation independent of T. As in the first example the index T on the left hand side is then left off.

If the density of the particle is close to that of the liquid or if it is small enough it will not sink to the bottom and one has a stationary process. The integral is then not dependent on the time one starts the experiment.

In almost all physical experiments, one will generally not measure the required quantity as a continuous function of time but rather as a discrete series of values, S_i, where i is the index of the value measured at time t_i. If the time interval between two measurements is constant and if it is sufficiently small compared to the time scale on which the quantity S varies, one essentially can replace the time integration by a sum in the following way:

$$\frac{1}{T} \int_{t_o}^{t_o+T} S(t) dt \Longrightarrow \frac{1}{N} \sum_{i=1}^{N} S_i \tag{3.4}$$

We have assumed here that the first measurement was performed at time t_o and the last one at time $t_o + T$.

Ensemble Average. The temporal averaging scheme is the usual way an experimentalist obtains mean values. For a theoretician, however, this averaging scheme is hardly one he can actually apply. The reason for that is fairly obvious: If he were to calculate a temporal average, he would need to calculate the temporal evolution of the respective quantity. Hence, for the case of Brownian motion of a particle in the fluid, he would need to calculate the motion of all particles within that fluid, because the motion of all these particles are coupled. With about 10^{23} in a cubic centimetre this is an entirely impossible task.

In the last century, the American physicist Gibbs had the idea to proceed as follows: One could consider an 'ensemble' of 'macroscopically identical' systems. 'Macroscopically identical' means that all macroscopic variables, such as composition, mass, pressure, temperature and dimensions of the vessels, are identical. If some of these quantities are dependent on time or space variables, they should be so in all of the systems. Let us now assume that for each system, i, a researcher measures the quantity $S_i^2(t;\tau)$ of a particle. With the set of these values one can calculate the following average value

$$\langle S^2(t;\tau)\rangle = \frac{1}{N}\sum_{i=1}^{N} S_i^2(t;\tau), \tag{3.5}$$

which is called the ensemble average.

It is by no means immediately obvious that this ensemble average is, for stationary processes, identical to the time average. The time average is a value obtained from measurements of *one* particle, whereas for the ensemble average we use values measured on different particles in different systems. The assumption that these averages are equal was extremely fruitful in the development of statistical mechanics, that is in the theory of systems with an extremely large number of particles. If the two averages are actually equal one calls the respective stochastic process 'ergodic'. It is, in general, not possible to prove the ergodicity for a real physical process. So there exists no real proof that for a Brownian particle the two averages are identical. However, one usually assumes ergodicity if there is no hint of 'non-ergodic' behaviour.

The mean intensity can also be formed by an ensemble averaging procedure. For that purpose one imagines a large number N of macroscopically identical light scattering set-ups. Again, one assumes that at a particular instant the researchers in each of these experiments measure the light intensity at a certain point on the observation screen. From these N measured values one can then calculate the ensemble average of the light intensity:

$$\langle I(\boldsymbol{r},t)\rangle = \frac{1}{N}\sum_{i=1}^{N} I_i(\boldsymbol{r},t) \tag{3.6}$$

It is again assumed that this value is the same as the one obtained through 'temporal averaging'.

The enormous advantage of ensemble averages is the fact that these can be calculated theoretically if some plausible assumptions are made concerning certain properties of the considered system. The way in which ensemble average values are calculated will be considered more closely in the next chapter. Before we familiarize ourselves with that procedure, we will look at another procedure to calculate average values, which, however, can only be used in cases of spatially dependent stochastic processes.

Spatial Average. Spatially dependent stochastic processes can be averaged over certain regions in space. (Spatial in this context denotes dependence on a space variable; it does not mean three dimensional. So, for instance, is the light intensity on a screen a spatial stochastic process even though it is a two dimensional one). It is possible to calculate the average intensity on the screen by a spatial integration in the following way:

$$\overline{I(\boldsymbol{r},t)}^{A} = \frac{1}{A} \cdot \int_{A} I(\boldsymbol{r},t) dA, \tag{3.7}$$

Here dA denotes the elements of the area A and $\boldsymbol{r}$ the vectors in the plane of the screen.

If the integration area is chosen considerably smaller than the screen area one can investigate whether the average value according to equation (3.7) is dependent on the region where the integration on the screen is located. If the average value turns out to be independent of the choice of this area, the intensity is called a homogeneous stochastic process.

The homogeneity of the light intensity pattern can, in the considered case, also be investigated by temporal averaging, provided this stochastic process is stationary. Then one can measure temporal averages at different locations and check whether these averages are equal or not.

Homogeneity of a stochastic process is the spatial analogy to the temporal property of stationarity. If furthermore a stochastic process is independent of direction it is called isotropic. .

Spatial stochastic processes are usually not homogeneous in a very rigorous sense. Any average values depend, even if only slightly, on the location where the averaging is performed. It is therefore advisable to check whether the homogeneity is sufficient when spatial averaging is performed.

3.3 Probabilities

3.3.1 The Probability Density

The expression (3.5) for ensemble averages can be reformulated to obtain an expression, which is much more suited for theoretical treatments. We will perform that reformulation using the example of the intensity measurement. For that purpose we divide the range of possible values for I, i.e. $(0 \leq I \leq \infty)$, into intervals dI_ν. Then any measured value I_i of the intensity belongs to a definite interval $[I_\nu, I_\nu + dI_\nu]$, where $I_\nu \leq I_i < I_\nu + dI_\nu$. Now let us assume that all values of a measurement set are distributed into these intervals, such that in the ν-th interval are $N(I_\nu)$ values. For the following considerations it is helpful to introduce the quantity

$$n(I_\nu) = \frac{N(I_\nu)}{dI_\nu}, \tag{3.8}$$

which describes the fraction of values falling into the ν-th interval.

The average value (3.6) can now be written approximately as

$$\langle I \rangle \approx \frac{1}{N} \sum_{\nu} I_\nu \cdot n(I_\nu) dI_\nu, \tag{3.9}$$

The sum is now extended over all intervals instead of all measured values. Since every measured value belongs to exactly one interval in the sum, all values appear exactly once. The only difference between the two formulae (3.6) and (3.9) is due to the fact, that in the second case the sum does not contain the actually measured values but the representatives of the respective intervals. This error can be made smaller if the width of the intervals is reduced.

This reformulation leads to a change in the interpretation of the two formulae: In (3.9) the quantity I_ν represents no longer a measured value, but it is determined by the lower limit of the respective interval. Thus, its value can be chosen to be the same in completely different experiments. The information on the measured values is now contained in the quantity $n(I_\nu)$, which tells us how many values have a certain size.

It is now possible to slightly rewrite the last formula in order to replace the summation by an integration: The summation over the intervals dI_ν can be expressed in the following way:

$$\langle I \rangle \approx \frac{1}{N} \sum_{dI} I \cdot n(I) dI. \tag{3.10}$$

One can now define the so called probability density P:

$$P^1(I) = \frac{n(I)}{N} \tag{3.11}$$

The superscript 1 denotes the dependence of this probability on one stochastic variable only. The transition to infinitesimal interval size dI then finally yields:

$$\langle I \rangle = \int_0^\infty P^1(I) \cdot I \cdot dI \tag{3.12}$$

The reader should not consider the reasoning just exposed as a derivation. The intention was rather to give an intuitive understanding of the notion probability density.

Since, in the case considered here, the intensity on the screen is a function of time and space, the probability density will also be a function of these variables. The last equation can then be written in the form

$$\langle I(\boldsymbol{r},t)\rangle = \int_0^\infty P^1(I;\boldsymbol{r},t)\cdot I\cdot dI \tag{3.13}$$

The function $P^1(I;\boldsymbol{r},t)$ is a 'true' function of the independent variable I; there is no way to see whether any stochastic property is connected to this function. It is only the interpretation of the function $P^1(I;\boldsymbol{r},t)$ as a probability density, which states that this function describes properties of a stochastic variable. To express this fact, the variable I is called a stochastic variable. That this function depends on one stochastic variable only is denoted by the superscript 1. The fact that in our case the stochastic quantity represents a time and space dependent process is expressed by the dependence of P^1 on $\boldsymbol{r}$ and t. Note that in this description the variable I is no longer itself a function of $\boldsymbol{r}$ and t. These variables, on which this function is dependent, are called parameters. The 'true' stochastic variables will usually be separated from the parameters by a semicolon, a notation that was also used in equation (3.13).

P^1 is subject to the following important normalization condition

$$\int_0^\infty P^1(I;\boldsymbol{r},t)dI = 1, \tag{3.14}$$

which expresses the fact that it is absolutely certain - i.e. the probability is 1 - that any measured value falls into the interval $[0,\infty]$.

The function P^1 is *no* probability, even though it is often called one. In our case this can be seen from the fact that it has the dimension $intensity^{-1}$. Only the expression $P^1(I;\boldsymbol{r},t)dI$ is a probability, namely the probability for an arbitrarily measured value to lie within the interval $[I,I+dI]$. Then the probability that a measured value is exactly equal to a value determined prior to the measurement is zero, because in that case $dI=0$. (This does not mean that this is impossible, it just means that it happens 'almost never').

If it is not the average I itself that is of interest but that of some function of I, $f(I)$, one has

$$\langle f(I(\boldsymbol{r},t))\rangle = \int_0^\infty P^1(I;\boldsymbol{r},t)\cdot f(I)\cdot dI \tag{3.15}$$

The dependence of this average on $\boldsymbol{r}$ and t is noted on the left hand side in the usual way. So the left hand side is a true function of $\boldsymbol{r}$ and t; there is no randomness combined with it.

3.3.2 Joint Probabilities

With the concept of the probability density function, it is now possible to tackle even more complicated problems like, for example, the calculation of the average value of the product $I(\boldsymbol{r}_2, t_1) \cdot I(\boldsymbol{r}_2, t_2)$. For this calculation, a more complex probability density function is needed, which gives the probability that at time t_1 and location $\boldsymbol{r}_1$, the light intensity is δI_1, and at time t_2 and location $\boldsymbol{r}_2$ it is δI_2. The phrase 'the light intensity is δI_1' means that the light intensity lies in the interval $[I_1, I_1 + dI_1]$. The desired probability density is a function of two random variables I_1 and I_2, and it also depends parametrically on the two pairs of space and time variables $\boldsymbol{r}_1, t_1$ and $\boldsymbol{r}_2, t_2$:

$$P^2(I_1, I_2; \boldsymbol{r}_1, \boldsymbol{r}_2, t_1, t_2) dI_1 dI_2 .$$

A probability density, which depends on two (or three ..) random variables, is called a joint probability density of second (third ...) order. In this book, the order of the probability density is always indicated by a superscript, as just shown for the probability density of second order.

If a probability density of second order is integrated over one of the random variables one obtains the probability density function of first order:

$$P^1(I_1; \boldsymbol{r}_1, t_1) = \int_0^\infty P^2(I_1, I_2; \boldsymbol{r}_1, \boldsymbol{r}_2, t_1, t_2) dI_2 \tag{3.16}$$

The function on the left hand side is just the ordinary probability density of equation (3.11), only with other symbols for the variables.

It is worthwhile noting that the integration on the right hand side of the last equation automatically removes the dependence on the parameters $\boldsymbol{r}_2$ and t_2 from the remaining function. In the same way as we integrated the second order probability density over I_2 we could have carried out the integration over I_1. The results would be exactly the same apart from the symbols of the variables.

A particularly simple form of joint probability densities occurs if some of the random variables are mutually independent. Variables are said to be independent, if the probability for one variable to assume a certain value within an arbitrarily chosen interval is independent of the value of the other variable. That means that the joint probability density of second order decomposes into a product of two probabilities of first order, if the two variables are independent. We will write that property down explicitly for the case of two intensities on a screen:

$$P^2(I_1, I_2; \boldsymbol{r}_1, \boldsymbol{r}_2, t_1, t_2) = P^1(I_1; \boldsymbol{r}_1, t_1) \cdot P^1(I_2; \boldsymbol{r}_2, t_2). \tag{3.17}$$

As will be seen in Sect. 4.6, this equation is true when the two locations $\boldsymbol{r}_1$ and $\boldsymbol{r}_2$ are sufficiently far apart.

For higher order probabilities and groups of independent variables, the decomposition of the probability density into products is analogous.

3.3.3 Conditional Probabilities

It is often helpful to use still another type of probability functions, the so called conditional probabilities. They answer the following question: What is the probability for one random variable to assume a certain value provided - or: under the condition - that the value of the second variable has some definite value. In case of the light intensity, one could be interested in the intensity I at location $\boldsymbol{r}$ and time t only for those situations in which, at a fixed time t_1 and fixed location $\boldsymbol{r}_1$, the exact intensity I_1 occurs. The usually adopted notation for this probability is

$$P_c(I; \boldsymbol{r}, t | I_1; \boldsymbol{r}_1, t_1) dI$$

Although the variable I_1 is in this case *not* a random variable in the strict sense, it is separated from the parameters by a semicolon.

Conditional probabilities can easily be expressed by simple expressions using 'normal' joint probabilities. So we have for instance

$$\begin{aligned} P^2(I_1, I_2; \boldsymbol{r}_1, \boldsymbol{r}_2, t_1, t_2) dI_1 dI_2 = \; & P^1(I_1; \boldsymbol{r}_1, t_1) dI_1 \cdot \\ & P_c(I_2; \boldsymbol{r}_2, t_2 | I_1; \boldsymbol{r}_1, t_1) dI_2 \end{aligned} \tag{3.18}$$

The right hand side is the probability, that the intensity at $\boldsymbol{r}_1$ and t_1 is found in the interval $[I_1, I_1 + dI_1]$, multiplied by the probability, that the intensity at $\boldsymbol{r}_2$ and t_2 lies in the interval $[I_2, I_2 + dI_2]$, provided the intensity has the value I_1 at $\boldsymbol{r}_1$ at time t_1. This, however, is exactly the meaning of the joint probability on the left hand side, it is simply expressed a bit more complicated.

Interchanging the variables and parameters on the right hand side leads to an equally valid relation:

$$\begin{aligned} P^2(I_1, I_2; \boldsymbol{r}_1, \boldsymbol{r}_2, t_1, t_2) dI_1 dI_2 = \; & P^1(I_2; \boldsymbol{r}_2, t_2) dI_2 \cdot \\ & P_c(I_1; \boldsymbol{r}_1, t_1 | I_2; \boldsymbol{r}_2, t_2) dI_1 \end{aligned} \tag{3.19}$$

3.3.4 Some Properties of Probability Densities

Transformation of Variables. It is sometimes a considerable advantage to introduce a new stochastic variable, which is connected to the old one by some transformation. The procedure to obtain the probability densities for the new variable is very much the same as that adopted for the transformations of variables in the calculation of integrals. If $P^1(I)$ is the probability density of the stochastic variable I, and if the new variable S is connected to I by $S = S(I)$, one has

$$\int_a^b P^1(I) dI = \int_{I(a)}^{I(b)} P^1(I(S)) \cdot \frac{dI(S)}{dS} \cdot dS, \tag{3.20}$$

where we have assumed that the transformation can be inverted: $I = I(S)$. Then the function $P^1(I(S)) \cdot \frac{dI(S)}{dS}$ is the probability density of the new variable S. Any parameters, on which $P^1(I)$ might also be dependent, are not influenced by this transformation. As an example, let us consider the probability density of the form

$$P^1(I) = \sqrt{\frac{\pi}{\alpha}} e^{-\alpha(I-I_o)^2}. \tag{3.21}$$

If, for some reason, one prefers to consider the stochastic variable $S = (I - I_o)^2$, the inverse transformation gives

$$I(S) = \sqrt{S} + I_o, \tag{3.22}$$

and the derivative

$$\frac{dI}{dS} = \frac{1}{2\sqrt{S} + I_o}$$

gives the new probability density

$$P^1(S) = \frac{1}{2}\sqrt{\frac{\pi}{\alpha}} \frac{1}{\sqrt{S}} e^{-\alpha S} \tag{3.23}$$

Stationarity and Homogeneity. Probability densities for stationary and/-or homogeneous processes have special properties, which in many cases simplify the calculation of average values considerably. We argued in Sect. 3.3.1 that probability densities can be looked at as ensemble averages. A more rigorous definition of the term stationarity as the one given above (see discussion preceding (3.3)) is that average values must not depend on the - arbitrarily - chosen zero of the time scale. This means that an arbitrary amount τ can be added to all times $t_1, t_2, ..t_n$, on which an average depends, without altering the value of the average. I.e. for a joint probability density of the second order we have

$$P^2(I_1, I_2; \boldsymbol{r}_1, \boldsymbol{r}_2, t_1, t_2) = P^2(I_1, I_2; \boldsymbol{r}_1, \boldsymbol{r}_2, t_1 + \tau, t_2 + \tau)$$

Choosing $\tau = -t_2$ yields

$$P^2(I_1, I_2; \boldsymbol{r}_1, \boldsymbol{r}_2, t_1, t_2) = P^2(I_1, I_2; \boldsymbol{r}_1, \boldsymbol{r}_2, t_1 - t_2, 0)$$

If the considered process is also homogeneous, one has the analogue property with respect to the space variable $\boldsymbol{r}$:

$$P^2(I_1, I_2; \boldsymbol{r}_1, \boldsymbol{r}_2; t_1, t_2) = P^2(I_1, I_2; \boldsymbol{r}_1 - \boldsymbol{r}_2, 0, t_1 - t_2, 0).$$

Usually the zeros are skipped and one writes $P^2(I_1, I_2; \boldsymbol{r}, t)$, where now the variables $\boldsymbol{r}$ and t are understood as space and time differences, respectively.

From the above, we can immediately conclude that the probability density of first order is entirely independent of space and time variables.

If, furthermore, the considered process is isotropic, the considered probability density is only dependent on the absolute value of $\boldsymbol{r}$:

$$P^2(I_1, I_2; |\boldsymbol{r}_1 - \boldsymbol{r}_2|, t_1 - t_2)$$

3.3.5 Particles in a Vessel

At the end of this chapter we will consider once again the motion of particles suspended in a fluid. Let us denote the probability to find a definite particle at time t in a volume dV at $\boldsymbol{r}$ by $P^1(\boldsymbol{r},t)dV$. Due to the fact that a particle can only be at one particular location in space, the variable $\boldsymbol{r}$ can be considered as the stochastic variable as well as a parameter. Therefore, it can assume all possible values, which denote locations within the considered vessel, and at the same time the normalization condition applies:

$$\int_V P^1(\boldsymbol{r};t)dV = 1, \tag{3.24}$$

since the probability to find the particle somewhere in the vessel is 1.

Again, one can define joint and conditional probabilities. Thus, for the probability to find a definite particle at time t_1 at location $\boldsymbol{r}_1$ in the volume dV_1, and at time t_2 in the volume dV_2 at $\boldsymbol{r}_2$, one has the joint probability

$$P^2(\boldsymbol{r}_1,\boldsymbol{r}_2;t_1,t_2)dV_1dV_2$$

Homogeneity means in this case that this probability is subject to the following relation:

$$P^2(\boldsymbol{r}_1,\boldsymbol{r}_2;t_1,t_2)dV_1dV_2 = P^2(\boldsymbol{r}_1-\boldsymbol{r}_2,0;t_1,t_2)dV_1dV_2 \tag{3.25}$$

Furthermore, if the process is stationary, one has

$$P^2(\boldsymbol{r}_1,\boldsymbol{r}_2;t_1,t_2)dV_1dV_2 = P^2(\boldsymbol{r}_1-\boldsymbol{r}_2,\boldsymbol{0};t_1-t_2,0)dV_1dV_2 \tag{3.26}$$

This notation indicates that the homogeneity refers to the stochastic variable itself, which is in contrast to the examples discussed above. It is now possible to express for this case the joint probability by use of equation (3.18) in the following way by the conditional probability:

$$\begin{aligned} &P^2(\boldsymbol{r}_1,\boldsymbol{r}_2;t_1,t_2)dV_1dV_2 \\ &= P^1(\boldsymbol{0},0)\cdot P_c(\boldsymbol{r}_1-\boldsymbol{r}_2,t_1-t_2|\boldsymbol{0},0)dV_1dV_2 \\ &= \frac{1}{V}P_c(\boldsymbol{r}_1-\boldsymbol{r}_2,t_1-t_2|\boldsymbol{0},0)dV_{12} \end{aligned} \tag{3.27}$$

where we have used that $P^1(\boldsymbol{0},0) = 1/V$, and dV_{12} is the volume element of the variable $\boldsymbol{r}_1-\boldsymbol{r}_2$.

In light scattering experiments, very many more or less identical particles are usually present in the considered fluid. For the definition of the joint probability it is therefore generally not useful to ask for the *same* particle in a certain region at a later time, but whether there is *any* particle at all. The probability function defined in this way is therefore the sum of the probability, that a definite particle moves from one location to another, and the probability, that the two particles are different ones. If the average distance of the particles is large compared to their diameter, it is a good approximation to consider the second contribution to the probability to be a constant.

3.4 Correlation Functions

As we noted in Sect. 3.2, it is in many cases possible to make statements about stochastic processes on the basis of plausibility considerations. Whether these statements are actually true, can usually not be proven within the framework of the theory. It is therefore necessary to compare theoretical and experimental results.

The most desirable quantities to measure were the probability densities. This, however, is a very hard task. Even for a simple time dependent stochastic process, the most simple probability density, which tells something about temporal behaviour, is the joint probability of second order. This is already a function of three variables: the two stochastic variables and time.

There is a much simpler function which contains information on the temporal behaviour but depends only on time: The correlation function. Let us consider as an example the stationary stochastic process $S(t)$. Then

$$C_S(\tau) = \langle S(t)S(t+\tau)\rangle \tag{3.28}$$

is called the correlation function of the quantity S. The more precise name is auto-correlation function, since the two factors on the right hand side represent the same stochastic process. If the two factors represent different processes one calls this function a cross-correlation function. A theoretical calculation of the correlation function requires the knowledge of the joint probability density of second order:

$$C_S(\tau) = \int\int P^2(S, S'; \tau) S S' dS dS'. \tag{3.29}$$

For non-stationary processes the correlation function evidently depends on two times and not just on the time difference.

Correlation functions have the big advantages that

- they can easily be calculated or modeled theoretically
- they can, in most cases, be measured without too much difficulty
- they are the most simple average values that say something about the time behaviour of a stochastic process.

But the correlation function also has a disadvantage: There is no direct intuitive way to interpret it. For that reason it usually takes some time to get familiar with correlation functions in order to extract the information they carry. We will therefore discuss a few properties which are common to all correlation functions.

All auto-correlation functions have an absolute maximum at the origin, i.e. when the value of the variables, on which the function depends, is zero. In case of a simple time dependent auto-correlation function one therefore has:

$$C_S(0) \geq C_S(\tau) \tag{3.30}$$

for arbitrary τ. This can be proven in the following way: From the obviously true relation

$$(S(0) - S(\tau))^2 \geq 0 \tag{3.31}$$

follows immediately

$$\left\langle (S(0) - S(\tau))^2 \right\rangle \geq 0 \tag{3.32}$$

Decomposition of the square yields

$$\langle S(0) \cdot S(0) \rangle - 2 \langle S(0) \cdot S(\tau) \rangle + \langle S(\tau) \cdot S(\tau) \rangle \geq 0. \tag{3.33}$$

We used here a property of the formation of an average: Summation - and also integration and differentiation - and formation of the average can be performed in an arbitrary order. The average of a sum is therefore identical to the sum of the corresponding averages.

Since the considered process is stationary the first term of the right hand side equals the last one. By use of equation (3.28), we then obtain

$$2C_S(0) - 2C_S(\tau) \geq 0 \tag{3.34}$$

or

$$C_S(0) \geq C_S(\tau) \tag{3.35}$$

A second property we are going to consider is the behaviour of the correlation function in the limit $\tau \to \infty$. Generally, the statistical fluctuations of a quantity S at two times will become more and more independent the larger the time interval between the times is. If that were not the case, one would not speak of random fluctuations. That means that for large enough τ, which we will denote by T, we can approximate the correlation function in the following way:

$$C_S(T) = \langle S(t) \cdot S(t+T) \rangle \approx \langle S(t) \rangle \cdot \langle S(t+T) \rangle = \langle S(0) \rangle^2 . \tag{3.36}$$

This property can be expressed exactly as follows:

$$\lim_{\tau \to \infty} C_S(\tau) = C_S(\infty) = \langle S(0) \rangle^2 . \tag{3.37}$$

Thus we can see that the correlation function drops from its largest value at $\tau = 0$ to the smaller value for $\tau = \infty$. The time scale of the fluctuations of the stochastic process affects the way in which the function decreases.

The variance $\Delta(S)$ of the quantity S, which is defined by the two values $C_S(\infty)$ and $C_S(0)$, can easily be expressed by

$$\Delta(S(t)) = \left\langle (S(t) - \langle S(t) \rangle)^2 \right\rangle = \left\langle S^2(t) \right\rangle - \langle S(t) \rangle^2 . \tag{3.38}$$

This can be shown by using equation (3.37) and writing

$$\left\langle S^2(t) \right\rangle - \langle S(t) \rangle^2 = \langle S(t) \cdot S(t) \rangle - \langle S(0) \rangle^2 = C_S(0) - C_S(\infty) \tag{3.39}$$

An intuitive access to the meaning of the correlation function is facilitated if the stochastic variable

$$s(t) = S(t) - \langle S(t) \rangle \tag{3.40}$$

is considered, which describes the deviations of the stochastic variable from its average value. One immediately realizes the following connection of the two correlation functions $C_s(\tau)$ and $C_S(\tau)$:

$$C_s(\tau) = C_S(\tau) - \langle S(0) \rangle^2 . \tag{3.41}$$

Together with equation (3.37) we then obtain

$$C_s(\infty) = 0. \tag{3.42}$$

Figure 3.1 represents an example of the formation of the correlation function, which is designed to get an intuitive understanding of the meaning of the correlation function. A stochastic artificial signal was generated with a computer. It is shown in part a) of the figure. The abscissa represents time in arbitrary units. The solid graphs in parts b) through d) are identical to this signal. In addition to this, time shifted signals are also shown as dotted graphs. The shift time τ is given in the respective part of the figure. The lower graphs represent the products $s(t)s(t + \tau)$. The time average of these products are the values of the correlation function for the respective τ. Apart from a constant factor, this time average is just the integral over the product function, which is nothing else than the area under the graph. It is seen in this figure that this area is biggest for $\tau = 0$ and decreases with increasing τ.

Part e) of the figure represents the correlation function calculated from the shown signal. The integration was carried out over a time interval, which was one hundred times longer than that shown in this figure. Note that the scale of the abscissa in part e) is increased by a factor of 10 with respect to the other parts. Furthermore, the quantity in this part is not the time itself but the time difference.

It has to be mentioned, that the correlation function need not be a function that decreases monotonously to its value $C_S(\infty)$. The correlation function shown in part e) of the Fig. 3.1, for instance, has negative values in the time interval $[\tau_1 \approx 40, \tau_2 \approx 70]$. Even though these negative values are very small in this case, it is a real effect due to the construction of this artificial stochastic process. The interpretation of this property is the following: If, at any time t, the value of the stochastic variable is larger (smaller) than the average value, then there is a large probability that in the interval $[t+\tau_1, t+\tau_2]$, the value of this variable is smaller (larger) than the average. The 'depressions' in the correlation function therefore express 'anti-correlations'. There are also correlation functions with several maxima and minima. The absolute value of these extrema can, however, never be larger than the maximum $C_S(0)$. The interpretation of these extrema is analogous to the above.

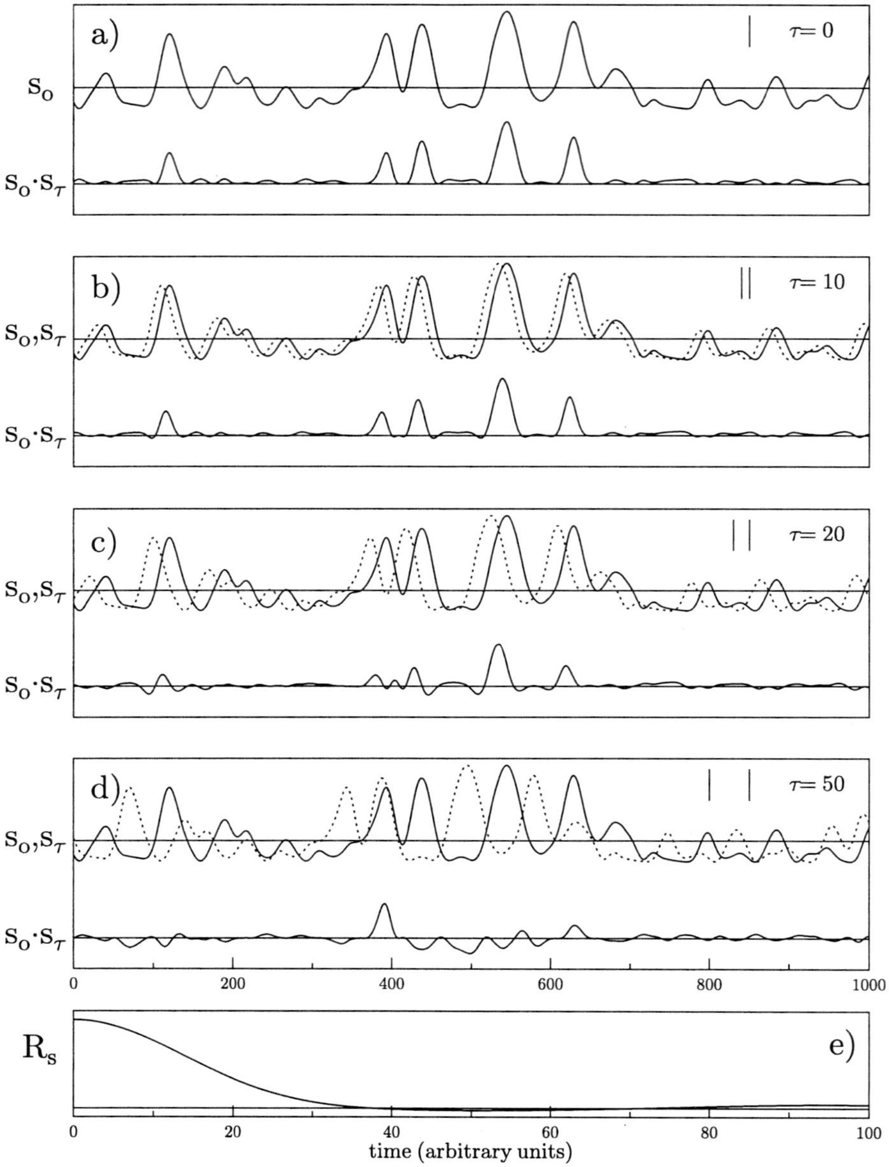

Fig. 3.1. For an intuitive approach to the meaning of the correlation function. Explanation in the text

A very rough characterization of a correlation function and thus of the underlying stochastic process can be made with the 'correlation time' τ_c. It is a parameter, which roughly denotes how fast the correlation function drops to its value $C_S(\infty)$. The meaning of this parameter is especially clear in cases where the correlation function is an exponential or a Gaussian function (two relatively common cases). One then has

$$C_S(\tau) = \exp\left(-\frac{|\tau|}{\tau_c}\right)\left(\langle S(0)^2\rangle - C_S(\infty)\right) + C_S(\infty) \tag{3.43}$$

and

$$C_S(\tau) = \exp\left(-\frac{\tau^2}{\tau_c^2}\right)\left(\langle S(0)^2\rangle - C_S(\infty)\right) + C_S(\infty), \tag{3.44}$$

respectively.

It is obvious, that the characterization of the correlation function with just one parameter is generally not sufficient in cases, where the correlation function has a more complex time dependence, for instance if there are 'depressions'. Nevertheless, the parameter 'correlation time' is widely used.

The correlation time τ_c can be interpreted as a measure of the 'memory' of the stochastic process. It specifies the time after which the process does no longer 'know' its former behaviour (or, putting it a bit more exactly, after which it does only know very little of its former behaviour).

A function that is directly connected to the correlation function, but which can be interpreted in an easier way, is the structure function $K_S(\tau)$:

$$\begin{aligned} K_S(\tau) &= \left\langle (S(t+\tau) - S(t))^2 \right\rangle \\ &= \left\langle S(t+\tau)^2 \right\rangle + \left\langle S(t)^2 \right\rangle - 2\left\langle (S(t+\tau)S(t)\right\rangle \\ &= 2\left\langle S(0)^2\right\rangle - 2C_S(\tau) = 2C_S(0) - 2C_S(\tau) \end{aligned} \tag{3.45}$$

The first line of this equation shows that the structure function is just the square deviation of the stochastic variable at two different times. It has obviously the properties $K_S(0) = 0$ and $K_S(\infty) = 2(C_S(0) - C_S(\infty))$.

What we said about the temporal behaviour of a time dependent process is mutatis mutandis true for a space dependent process. For instance, for such processes it is possible to define correlation lengths on the basis of the spatial correlation function. For processes which depend on one, two, or three spatial co-ordinates, one then has correlation lengths, areas, or volumes.

For space and time dependent processes, which are stationary and homogeneous, the correlation function depends on time and space differences:

$$C_S(\boldsymbol{\rho}, \tau) = \langle S(\boldsymbol{r} + \boldsymbol{\rho}, t + \tau) S(\boldsymbol{r}, t)\rangle \tag{3.46}$$

The maximum property (c.f. equation 3.30) for such correlation functions is then simply:

$$C_S(0,0) \geq C_S(\boldsymbol{\rho}, \tau) \tag{3.47}$$

3.5 The Velocity of Space and Time Dependent Stochastic Processes

It is generally possible to define the velocity of a space and time dependent stochastic process. We will consider here only a special type of process where the notion of the velocity of a stochastic process can be perceived very simply. Let us consider a process, which depends on $\boldsymbol{r}$ and t in the following way:

$$S(\boldsymbol{r}, t) = S(\boldsymbol{r} - \boldsymbol{v}t) \tag{3.48}$$

In the case that

- $S(\boldsymbol{r} - \boldsymbol{v}t) \geq 0$ and
- $\boldsymbol{r}$ is a vector in a plane,

the velocity of this process can be visualized by imagining $S(\boldsymbol{r}, t)$ to be the intensity of a light pattern on a screen. A process with the property (3.48) represents evidently a stochastic moving light pattern on the screen with the velocity $\boldsymbol{v}$.

Generally, the pattern can also change in a random way with time while it is moving. In that case, it is not possible to decompose the temporal change of the pattern at any location on the screen into a moving and a randomly changing contribution. For a temporal change at just one location can only be an increase or decrease of the intensity, and this can result from both mentioned types of pattern change. We will not investigate the general question under which conditions a velocity can be attributed to a pattern. We will only consider a type of stochastic pattern which obviously has such a velocity. This type is characterized by the following property: All joint probabilities depend only on the parameters $\boldsymbol{r_i}, t_i$ through $\boldsymbol{\xi}_i = \boldsymbol{r}_i - \boldsymbol{v}t_i$ and $|t_i|$.

As was shown above, the form of the variable $\boldsymbol{\xi}_i = \boldsymbol{r}_i - \boldsymbol{v}t_i$ expresses the motion of the pattern with a velocity $\boldsymbol{v}$, whereas the dependence on $|t_i|$ describes a stochastic temporal change of the pattern. If one moved with the same velocity as the pattern one would only observe this temporal change caused by the dependence on $|t_i|$, which, in laser light scattering, is called the 'boiling' of the pattern.

We will now consider which average value - or average values - can be used to express the velocity of the pattern. Since velocity is a spatio-temporal phenomenon, it seems plausible that the most simple average value, which tells something about the velocity, is the spatio-temporal correlation function defined in equation (3.46). For the pattern considered here, we can write

$$\begin{aligned} C(\boldsymbol{\rho}, \tau) &= \langle (S(\boldsymbol{r} + \boldsymbol{\rho}, t + \tau) S(\boldsymbol{r}, t) \rangle \\ &= \int S \cdot S' P^2(S, S'; \boldsymbol{\rho} - \boldsymbol{v}\tau, |\tau|) dS dS'. \end{aligned} \tag{3.49}$$

The second line of this equation shows, that the correlation function depends on the two variables $\boldsymbol{\rho} - \boldsymbol{v}\tau$ and $|\tau|$ in the same way as the joint probability. For simplicity reasons we therefore write this correlation function in the form $C(\boldsymbol{\rho} - \boldsymbol{v}\tau, |\tau|)$. C has, as every correlation function, an absolute maximum at $(\boldsymbol{\rho} = 0, \tau = 0)$. We now assume that the correlation function is differentiable everywhere. This means that it depends on $|\tau|$ via $|\tau|^2 = \tau^2$. We can now expand the correlation function in a Taylor series around the origin. Due to the maximum at that point, all first derivatives must disappear and we have

$$\begin{aligned}
C(\boldsymbol{\rho} - \boldsymbol{v}\tau, |\tau|) &= C(0,0) + \frac{1}{2}C_{xx}(0,0)(\rho_x - v_x\tau)^2 + \\
&\frac{1}{2}C_{yy}(0,0)(\rho_y - v_y\tau)^2 + \\
&+C_{xy}(0,0)(\rho_x - v_x\tau)(\rho_y - v_y\tau) + \frac{1}{2}\ddot{C}(0,0)\tau^2 + \\
&+\mathit{residue} \\
&= C(0,0) + \frac{1}{2}\ddot{C}(0,0)\tau^2 + \frac{1}{2}\boldsymbol{\xi}\hat{A}\boldsymbol{\xi} + \mathit{residue}
\end{aligned} \tag{3.50}$$

In this equation we used the relations

$$\boldsymbol{\xi} = \boldsymbol{\rho} - \boldsymbol{v}\tau \tag{3.51}$$

and

$$\hat{A} = \begin{pmatrix} C_{xx}(0,0) & C_{xy}(0,0) \\ C_{yx}(0,0) & C_{yy}(0,0) \end{pmatrix}. \tag{3.52}$$

residue denotes all terms with higher derivatives.

The indices of C denote the derivatives with respect to the denoted spatial variable. So $C_{xx}(0,0)$ is the second derivative of C with respect to x at the location $(\boldsymbol{\rho}, \tau) = (0,0)$.

Using equation (3.50), the following second derivatives can be calculated:

$$\begin{aligned}
\frac{\partial^2 C(0,0)}{\partial\rho_x\partial\tau} &= \frac{\partial}{\partial\tau}\left(\frac{1}{2}C_{xx}(0,0)\cdot 2\cdot(\rho_x - v_x\tau) + C_{xy}(0,0)\cdot(\rho_y - v_y\tau)\right) \\
&= -C_{xx}(0,0)v_x - C_{xy}(0,0)v_y
\end{aligned} \tag{3.53}$$

and

$$\frac{\partial^2 C(0,0)}{\partial\rho_x^2} = C_{xx}(0,0). \tag{3.54}$$

Clearly the term $-C_{xy}(0,0)v_y$ impedes the derivation of the simple relation

$$\frac{\dfrac{\partial^2 C(0,0)}{\partial\rho_x\partial\tau}}{\dfrac{\partial^2 C(0,0)}{\partial\rho_x^2}} = -v_x \tag{3.55}$$

and analogously

$$\frac{\dfrac{\partial^2 C(0,0)}{\partial \rho_y \partial \tau}}{\dfrac{\partial^2 C(0,0)}{\partial \rho_y^2}} = -v_y \qquad (3.56)$$

But one can always eliminate this term just by choosing a co-ordinate system x', y', which is rotated with respect to the original co-ordinate system. It is determined by the property

$$C_{x'y'}(0,0) = 0. \qquad (3.57)$$

Such a co-ordinate system can always be found. That is especially easy to see if the correlation function has a mirror symmetry about an axis, which we will call y′-axis. In that case, we obviously have $C_{x'}(x' = 0, y') = 0$ for all y', from which follows immediately equation (3.57). From here on, we will always consider the correlation function in such a co-ordinate system. For isotropic patterns the condition (3.57) is fulfilled in any co-ordinate system.

Differentiation of equation (3.49) with respect to ρ_x yields

$$C_x(\boldsymbol{\rho}, \tau) = \langle (S_x(\boldsymbol{r} + \boldsymbol{\rho}, t + \tau) S(\boldsymbol{r}, t) \rangle \,. \qquad (3.58)$$

Due to the assumed homogeneity, equation (3.49) can also be written as:

$$C(\boldsymbol{\rho}, \tau) = \langle (S(\boldsymbol{r}, t + \tau) S(\boldsymbol{r} - \boldsymbol{\rho}, t) \rangle \qquad (3.59)$$

We then obtain for the derivative of the correlation function

$$\begin{aligned} C_x(\boldsymbol{\rho}, \tau) &= - \langle (S(\boldsymbol{r}, t + \tau) S_x(\boldsymbol{r} - \boldsymbol{\rho}, t) \rangle \\ &= - \langle (S(\boldsymbol{r} + \boldsymbol{\rho}, t + \tau) S_x(\boldsymbol{r}, t) \rangle \end{aligned} \qquad (3.60)$$

Thus, the derivative of a correlation function can be expressed as a correlation function of differentiated quantities. For homogeneous processes one obtains the special relation

$$\langle (S_x(\boldsymbol{r} + \boldsymbol{\rho}, t + \tau) S(\boldsymbol{r}, t) \rangle = - \langle (S(\boldsymbol{r} + \boldsymbol{\rho}, t + \tau) S_x(\boldsymbol{r}, t) \rangle \,. \qquad (3.61)$$

An analogous relation is clearly true for temporal derivatives of the correlation function of stationary processes.

With the derived properties one can rewrite equations (3.55) and (3.56) in the following form:

$$\frac{\left\langle \dot{S}(0,0) \cdot S_x(0,0) \right\rangle}{\langle S_x(0.0) \cdot S_x(0.0) \rangle} = - \langle v_x \rangle \,. \qquad (3.62)$$

It is possible to prove that this last relationship is a special case of the more general property

$$\left\langle \dot{S}(0,0) \cdot \mathrm{f}(S_x(0,0)) \right\rangle = - \langle v_x \rangle \cdot \langle S_x(0.0) \cdot \mathrm{f}(S_x(0.0)) \rangle \qquad (3.63)$$

[10]. There is no limitation to the choice of the function f, although this equation can obviously only be resolved with respect to pattern velocity if the right hand side of the equation is not equal to zero. Equation (3.63) is valid under the same conditions as equation (3.55).

We will finally point to the fact that the second factor on the right hand side of the last equation does not depend on the pattern velocity, because it depends on the pattern intensity at *one* time only.

3.6 Some Properties of Tensors

3.6.1 Transformation Properties of Tensors

We assume that the reader is familiar with the definition and the basic properties of matrices, as for instance, the rules of multiplication. We will treat only those properties that are essential for the understanding of the following chapters. The matrices of main importance in this book are those of the dimension 3×3:

$$\hat{D} = \begin{pmatrix} D_{11} & D_{12} & D_{13} \\ D_{21} & D_{22} & D_{23} \\ D_{31} & D_{32} & D_{33} \end{pmatrix}. \tag{3.64}$$

The indices 1,2,3 denote in the following way the co-ordinate axes x,y,z:

$$1 \leftrightarrow x, 2 \leftrightarrow y, 3 \leftrightarrow z \tag{3.65}$$

We will adopt both notations in parallel.

Tensors are mathematical objects with particular transformation properties. Furthermore, tensors which are used in this book are so called second rank tensors and can be written in form of the matrix given in equation (3.64). Before we are going to consider how tensors behave if the co-ordinate system is changed we will look at the behaviour of vectors in such a situation.

Consider a vector $\boldsymbol{a}$ determined by the components (a_1, a_2, a_3) in a particular co-ordinate system S. In a system S', which is rotated with respect to S by an arbitrary angle around some axis, this vector has the components a'_1, a'_2, a'_3, which are generally obtained by the multiplication of $\boldsymbol{a}$ with an orthogonal matrix $\hat{O}$:

$$\boldsymbol{a}' = \hat{O} \cdot \boldsymbol{a}. \tag{3.66}$$

Written down explicitly one has

$$\begin{pmatrix} a'_1 \\ a'_2 \\ a'_3 \end{pmatrix} = \begin{pmatrix} O_{11} & O_{12} & O_{13} \\ O_{21} & O_{22} & O_{23} \\ O_{31} & O_{32} & O_{33} \end{pmatrix} \cdot \begin{pmatrix} a_1 \\ a_2 \\ a_3 \end{pmatrix} \tag{3.67}$$

$$= \begin{pmatrix} O_{11}a_1 + O_{12}a_2 + O_{13}a_3 \\ O_{21}a_1 + O_{22}a_2 + O_{23}a_3 \\ O_{31}a_1 + O_{32}a_2 + O_{33}a_3 \end{pmatrix}$$

A common way to write the last equation is

$$a'_i = \sum_j O_{ij} a_j \tag{3.68}$$

where the summation index - as in all equations henceforth - extends from 1 to 3. An illustration of this property of vectors is shown in Fig. 3.2 a). It is obvious that the two symbols $\boldsymbol{a}$ and $\boldsymbol{a}'$ denote the *same* vector. But its representation is different because of the fact that this vector is expressed in two different co-ordinate systems, which leads to different components in the two co-ordinate systems.

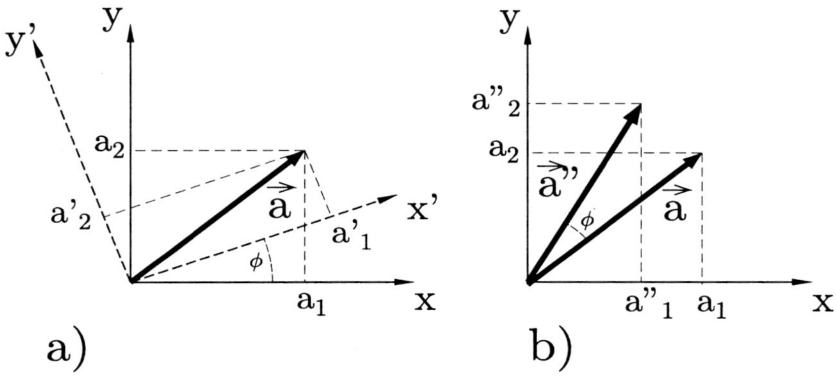

Fig. 3.2. The transformation of vectors by rotations. Explanation in the text.

Another situation is given if the co-ordinate system is kept fixed but the vector itself is rotated. This is shown in Fig. 3.2 b). The components of the rotated vector can again be obtained from the original one by multiplication with an orthogonal matrix in a way analogous to equation (3.66). We first note that the components of a vector of a given length only depend on the angles between the vector and the co-ordinate axes. Therefore, the components will be the same whether one rotates the co-ordinate system in one way or the vector in the opposite way. One can then conclude for the case shown in part b) of Fig. 3.2 that the following relation holds:

$$\boldsymbol{a}'' = \hat{O}^{-1} \cdot \boldsymbol{a}. \tag{3.69}$$

where $\hat{O}^{-1}$ is the matrix which describes the inverse rotation. A rotation of a vector back and forth apparently amounts to leaving the vector the same as before. That means that the multiplication of a vector with the matrix $\hat{O}$ and then with the matrix $\hat{O}^{-1}$ leaves the vector unchanged. So we have:

$$\hat{O}^{-1} \cdot \hat{O} = \hat{O} \cdot \hat{O}^{-1} = \hat{E}. \tag{3.70}$$

where $\hat{E}$ is the unit matrix

$$\hat{E} = \begin{pmatrix} 1\,0\,0 \\ 0\,1\,0 \\ 0\,0\,1 \end{pmatrix}. \tag{3.71}$$

The inverse of an orthogonal matrix is identical to its transpose:

$$\hat{O}^{-1} = \hat{O}^T. \tag{3.72}$$

This transpose $\hat{O}^T$ of a matrix $\hat{O}$ is obtained by just interchanging columns and rows of it:

$$O^T_{ij} = O_{ji}. \tag{3.73}$$

Using this property equation (3.69) can be solved with respect to $\boldsymbol{a}''$ by multiplying it on both sides with $\hat{O}^T$:

$$\hat{O}^T \cdot \hat{O} \cdot \boldsymbol{a}'' = \boldsymbol{a}'' = \hat{O}^T \cdot \boldsymbol{a} \tag{3.74}$$

The comparison of the equations (3.69) and (3.74) shows that any equation of this type can be interpreted either as a rotation of a vector or as a rotation of the co-ordinate system.

As an example, we will now write down explicitly the rotation matrices for the cases shown in Figs. 3.2 a) and 3.2b), respectively. The general form of a matrix for a rotation of the co-ordinate system around the z-axis about an angle ϕ (this corresponds to the case shown in part a) of the figure) is:

$$\hat{O}^c_z(\phi) = \begin{pmatrix} \cos(\phi) & \sin(\phi) & 0 \\ -\sin(\phi) & \cos(\phi) & 0 \\ 0 & 0 & 1 \end{pmatrix} \tag{3.75}$$

The superscript c denotes matrices which describe the transformation of the components of a vector if the co-ordinate system is rotated, whereas the superscript v refers to matrices which describe the transformation of the components of a vector if this vector is rotated in a fixed co-ordinate system. The components of the rotation matrix corresponding to part b) of the figure are obtained from this one by just replacing ϕ by $-\phi$. This amounts to reversing the sign of the two off-diagonal elements of the matrix. The resulting matrix is, furthermore, just the transposed of the one shown in equation (3.75):

$$\hat{O}^v_z(\phi) = \hat{O}^c_z(-\phi) = \begin{pmatrix} \cos(\phi) & -\sin(\phi) & 0 \\ \sin(\phi) & \cos(\phi) & 0 \\ 0 & 0 & 1 \end{pmatrix} \tag{3.76}$$

Matrices for rotations around the x-axis are obtained from the one given in equation (3.75) by 'cyclic permutation' of the elements. That means the element O_{ij} is moved to the location $i+1, j+1$. If an index assumes the value 4 it must then be replaced by 1. This procedure leads to the following matrix for rotations around the x-axis:

$$\hat{O}_x^c(\phi) = \begin{pmatrix} 1 & 0 & 0 \\ 0 & \cos(\phi) & \sin(\phi) \\ 0 & -\sin(\phi) & \cos(\phi) \end{pmatrix} \tag{3.77}$$

It is then easy to obtain the matrix for rotations around the y-axis:

$$\hat{O}_y^c(\phi) = \begin{pmatrix} \cos(\phi) & 0 & -\sin(\phi) \\ 0 & 1 & 0 \\ \sin(\phi) & 0 & \cos(\phi) \end{pmatrix} \tag{3.78}$$

The matrices $\hat{O}_x^v(\phi)$ and $\hat{O}_y^v(\phi)$ can again be obtained by just transposing those of the last two equations.

The exact meaning of the phrase 'rotation by a certain angle around an axis' is the following: If one looks opposite to the direction of the rotation axis, i.e. towards the origin of the co-ordinate system, then the rotation about the given angle is in the mathematical positive sense, i.e. counter-clockwise. Consequently, the rotations in Fig. 3.2 are around the z-axis by a positive angle ϕ.

If one has successive rotations around arbitrary axes the matrix for this composed rotation is obtained by a multiplication of the corresponding rotation matrices. Let us consider as an example the case of first rotating a vector in a fixed co-ordinate system around the x-axis about an angle ϕ and then around the y-axis about an angle ψ. The components of the rotated vector are then given by

$$\boldsymbol{a}'' = \hat{O}_y^v(\psi) \cdot \hat{O}_x^v(\phi) \cdot \boldsymbol{a}. \tag{3.79}$$

Multiplication of the two matrices yields the rotation matrix of the combined rotation:

$$\boldsymbol{a}'' = \begin{pmatrix} \cos(\psi) & \sin(\psi)\sin(\phi) & \sin(\psi)\cos(\phi) \\ 0 & \cos(\phi) & -\sin(\phi) \\ -\sin(\psi) & \cos(\psi)\sin(\phi) & \cos(\psi)\cos(\phi) \end{pmatrix} \cdot \boldsymbol{a}. \tag{3.80}$$

It is not difficult to convince oneself that the result of successive rotations depends on the order in which these rotations are performed. Consequently, the same is then true for the corresponding matrices. This means that for two arbitrary matrices $\hat{A}$ and $\hat{B}$ one generally has

$$\hat{A} \cdot \hat{B} \neq \hat{B} \cdot \hat{A}. \tag{3.81}$$

Again, equation (3.79) can be interpreted as a combined rotation of a co-ordinate system. To avoid inconsistencies, let us, for an exact interpretation, rewrite this in the following form:

$$\boldsymbol{a}' = \hat{O}_y^c(\psi) \cdot \hat{O}_x^c(\phi) \cdot \boldsymbol{a}. \tag{3.82}$$

Now, the matrix describing the rotation around the y-axis acts on the vector $\hat{O}^c_x(\phi) \cdot \boldsymbol{a}$ which is already in a rotated co-ordinate system. Therefore, the matrix $\hat{O}^c_y(\psi)$ induces a rotation around the *new* y-axis. It is essential that this property of combined rotations of the co-ordinate system is kept in mind.

Having recalled these transformation properties of vectors we will now look at the behaviour of tensors if rotations are carried out, i.e. how the components of a tensor are altered if one considers the tensor in a rotated co-ordinate system. For that purpose we consider the scalar product of two arbitrary vectors $\boldsymbol{a}$ and $\boldsymbol{b}$:

$$\boldsymbol{a} \cdot \boldsymbol{b} = \sum_i a_i \cdot b_i \tag{3.83}$$

Obviously, this quantity cannot be dependent on the choice of the co-ordinate system, in which the vectors are expressed. This implies that the scalar product must be independent of rotations of the co-ordinate system:

$$\boldsymbol{a} \cdot \boldsymbol{b} = \boldsymbol{a}' \cdot \boldsymbol{b}' \tag{3.84}$$

A tensor $\hat{D}$ is defined by the fact that the expression $\boldsymbol{a} \cdot (\hat{D} \cdot \boldsymbol{b})$, (which is usually written just as $\boldsymbol{a} \cdot \hat{D} \cdot \boldsymbol{b}$), is independent of the co-ordinate system, and is, therefore, also a scalar. This condition is sufficient to calculate the desired transformation property of $\hat{D}$ in a unique way.

Before we are able to do that we have to mention the following relation, which is valid for an arbitrary matrix $\hat{D}$ and arbitrary vectors:

$$\left(\hat{D} \cdot \boldsymbol{a}\right) \cdot \boldsymbol{b} = \boldsymbol{a} \cdot \hat{D}^T \cdot \boldsymbol{b}, \tag{3.85}$$

This can easily be shown:

$$\left(\hat{D} \cdot \boldsymbol{a}\right) \cdot \boldsymbol{b} = \sum_{i,j} D_{ij} a_j b_i = \sum_{i,j} a_j D^T_{ji} b_i = \boldsymbol{a} \cdot (\hat{D}^T \cdot \boldsymbol{b}) = \boldsymbol{a} \cdot \hat{D}^T \cdot \boldsymbol{b}. \tag{3.86}$$

We can now say:

$$\boldsymbol{a}' \cdot \hat{D}' \cdot \boldsymbol{b}' = \boldsymbol{a} \cdot \hat{D} \cdot \boldsymbol{b}, \tag{3.87}$$

where $\hat{D}'$ denotes the tensor in the rotated co-ordinate system. Replacing the primed vectors in this equation with the help of equation (3.66) yields

$$\begin{aligned} \boldsymbol{a}' \cdot \left(\hat{D}' \cdot \boldsymbol{b}'\right) &= \left(\hat{O}^c \cdot \boldsymbol{a}\right) \cdot \left(\hat{D}' \cdot \left(\hat{O}^c \cdot \boldsymbol{b}\right)\right) \\ &= \boldsymbol{a} \cdot \left(\left(\hat{O}^c\right)^T \cdot \hat{D}' \cdot \hat{O}^c \cdot \boldsymbol{b}\right) \\ &= \boldsymbol{a} \cdot \hat{D} \cdot \boldsymbol{b} \\ &= \boldsymbol{a} \cdot \left(\hat{D} \cdot \boldsymbol{b}\right) \end{aligned} \tag{3.88}$$

Since this equation must be true for arbitrary vectors $\boldsymbol{a}$ and $\boldsymbol{b}$ the matrices on the right hand side in the second and the last line must be the same and we have

$$\left(\hat{O}^c\right)^T \cdot \hat{D}' \cdot \hat{O}^c = \hat{D} \tag{3.89}$$

or

$$\hat{D}' = \hat{O}^c \cdot \hat{D} \cdot \left(\hat{O}^c\right)^T . \tag{3.90}$$

If a tensor is rotated in a fixed co-ordinate system instead of rotating the co-ordinate system we obtain

$$\hat{D}'' = \left(\hat{O}^c\right)^T \cdot \hat{D} \cdot \hat{O}^c = \hat{O}^v \cdot \hat{D} \cdot \left(\hat{O}^v\right)^T . \tag{3.91}$$

3.6.2 Functions of Tensors

We used in this chapter the property of the product of two tensors to be again a tensor. It is therefore possible to build functions of tensors, if these functions can be represented by power series. For that reason, the exponential of a tensor can easily be written down:

$$\exp\left(\hat{D}\right) = \mathrm{e}^{\hat{D}} = \hat{E} + \hat{D} + \frac{1}{2!}\hat{D}^2 + \frac{1}{3!}\hat{D}^3 + ... \tag{3.92}$$

A special property of the exponential of a tensor refers to traceless tensors. The trace of a tensor $\hat{D}$ is the sum of the diagonal elements:

$$\mathrm{Trace}\left(\hat{D}\right) = \sum_{i=1}^{3} D_{ii} \tag{3.93}$$

For a traceless tensor $\hat{S}$ we then have

$$\mathrm{Trace}\left(\hat{S}\right) = S_{xx} + S_{yy} + S_{zz} = 0. \tag{3.94}$$

Such a traceless tensor has the following property:
The tensor $\exp(-\hat{S})$ is the inverse of the tensor $\exp(\hat{S})$, i.e.

$$\exp(\hat{S}) \cdot \exp(-\hat{S}) = \exp(-\hat{S}) \cdot \exp(\hat{S}) = \hat{E} \tag{3.95}$$

4. Physical Basis: Theoretical Part

4.1 Introduction

A theoretical description of the phenomena which are the basis of the measurement techniques we describe here must deal with two different areas of physics. Firstly, the motion of Brownian particles in a fluid flow must be described. Secondly, we have to deal with stochastic light fields, which contain the desired information on the flow. In this chapter we are therefore going to consider the basis of these two areas in so far as they are necessary in order to understand the different measurement methods.

4.2 The Velocity Gradient

The velocity in a flowing fluid is a function of space and time: $\boldsymbol{u} = \boldsymbol{u}(\boldsymbol{r}, t)$. The nine quantities

$$\frac{\partial u_i}{\partial x_j} = \Gamma_{ij} \, (i, j = 1, 2, 3) \tag{4.1}$$

form the so called velocity gradient tensor, which is generally also a function of space and time. (For the notation of the components see equation (3.65)). In a real flow it is always possible to consider a small volume around a point of interest $\boldsymbol{r}$, in which this tensor can be viewed as practically independent of space. Within such a volume, which we will call volume of homogeneity, the velocity gradient is only dependent on time. Within the volume of homogeneity the velocity field of the fluid is given by:

$$\boldsymbol{u}(\boldsymbol{r}, t) = \boldsymbol{u}_o(t) + \hat{\Gamma}(t) \cdot \boldsymbol{r} \tag{4.2}$$

In this notation the size of $\boldsymbol{u}_o$ depends on the choice of the origin of the co-ordinate system, which will preferably be somewhere within this volume. The velocity gradient $\hat{\Gamma}$ is entirely independent of this choice. To begin with, the time dependence of the two quantities $\boldsymbol{u}_o$ and $\hat{\Gamma}$ will be neglected, i.e. we will first consider stationary laminar flows.

In many cases a fluid can be regarded as incompressible. Then we have

$$\mathrm{div}(\boldsymbol{u}) = \frac{\partial u_x}{\partial x} + \frac{\partial u_y}{\partial y} + \frac{\partial u_z}{\partial z} = \Gamma_{11} + \Gamma_{22} + \Gamma_{33} = 0 \tag{4.3}$$

or

$$\mathrm{Trace}(\hat{\Gamma}) = 0. \tag{4.4}$$

Thus, for incompressible fluids, which are of interest in this book, only 8 of the 9 components of the velocity gradient tensor $\hat{\Gamma}$ are linearly independent.

Every tensor can be separated into a symmetric and an antisymmetric part:

$$\hat{\Gamma} = \hat{\Gamma}^s + \hat{\Gamma}^a \tag{4.5}$$

where

$$\Gamma^s_{ij} = \Gamma^s_{ji} \text{ and } \Gamma^a_{ij} = -\Gamma^a_{ji} \tag{4.6}$$

By use of the transpose of the tensor one can simply express these two parts:

$$\begin{aligned} \hat{\Gamma}^s &= \frac{1}{2}\left(\hat{\Gamma} + \hat{\Gamma}^T\right) \\ \hat{\Gamma}^a &= \frac{1}{2}\left(\hat{\Gamma} - \hat{\Gamma}^T\right) \end{aligned} \tag{4.7}$$

The symmetric part of the velocity gradient tensor with its five components (six for compressible fluids) describes shear, whereas the three components of the antisymmetric part describe a rotation. The components of this latter part are simply connected with the rotation $\mathrm{rot}(\boldsymbol{u})$ of the fluid velocity:

$$\mathrm{rot}(\boldsymbol{u}) = 2(-\Gamma^a_{23}, \Gamma^a_{13}, -\Gamma^a_{12}) \tag{4.8}$$

or, more general

$$\mathrm{rot}(\boldsymbol{u}) = (\Gamma_{32} - \Gamma_{23}, \Gamma_{13} - \Gamma_{31}, \Gamma_{21} - \Gamma_{12}) \tag{4.9}$$

In fluid dynamics the quantity $\mathrm{rot}(\boldsymbol{u})$ is called vorticity.

4.3 Motion of Particles in a Homogeneous Velocity Field

As discussed in Sect. 2.3.1 optical measurement methods generally need light scattering particles immersed in the fluid under investigation. It is obvious that the calculation of the temporal and spatial behaviour of the scattered light requires the knowledge of the paths of the light scattering particles. Due to their stochastic motion one can at best have probabilities, as for instance the conditional probability to find a particle at time t within the infinitesimal volume element dV, provided it was at the location $\boldsymbol{r}_o$ at time t_o. Clearly this probability will not allow for the calculation of all properties of the scattered light field that can be looked at, but we will see that it is sufficient to calculate what is of interest here.

It is obvious that this conditional probability only depends on the time difference $t - t_o$, i.e. the considered process is stationary. We now perform the replacement $t - t_o \to t$. Therefore, the time variable t denotes in the following a time difference rather than an absolute time.

The conditional probability needed was denoted in Sect. 3.3 by the symbol $P_c(\boldsymbol{r}, t|\boldsymbol{r}_o, 0)dV$. The vector $\boldsymbol{r}$ is here the stochastic variable. For general velocity fields this probability can at best be calculated numerically. But for Brownian particles in a homogeneous velocity field independent of time it can be calculated. According to [11], the analytical form of the probability density is

$$P_c(\boldsymbol{r}, t|\boldsymbol{r}_o, 0) = \frac{1}{(4\pi D)^{3/2}\sqrt{|\hat{A}(t)|}} \cdot \tag{4.10}$$

$$\exp\left(-\frac{1}{4D}(\boldsymbol{r} - \hat{R}(t)\boldsymbol{r}_o - \hat{S}(t)\boldsymbol{u}_o) \cdot \hat{A}^{-1}(t)(\boldsymbol{r} - \hat{R}(t)\boldsymbol{r}_o - \hat{S}(t)\boldsymbol{u}_o)\right).$$

Here we used the following symbols:

$$\hat{R}(t) = \exp(\hat{\Gamma}t)$$

$$\hat{S}(t) = \int_0^t \exp(\Gamma t')dt'$$

$$\hat{A}(t) = \int_0^t \exp(\Gamma t')\exp(\Gamma^T t')dt'.$$

D is the diffusion constant of the particles and $|\hat{A}(t)|$ is the determinant of the matrix $\hat{A}(t)$.

Although this probability looks rather difficult, its basic structure is fairly simple. The matrices $\hat{A}(t)$, $\hat{R}(t)$, and $\hat{S}(t)$ are 3×3 tensors with time dependent components. The quantities $\hat{R}(t)\boldsymbol{r}_o$ and $\hat{S}(t)\boldsymbol{u}_o$ are therefore simply time dependent vectors.

As to its spatial dependence, the function in equation (4.10) is a three-dimensional Gaussian which, at time t, is centered around the location

$$\boldsymbol{r}(t) = \hat{R}(t) \cdot \boldsymbol{r}_o + \hat{S}(t) \cdot \boldsymbol{u}_o. \tag{4.11}$$

The extension of this Gaussian in space is determined by the tensor $A^{-1}(t)$. Areas of constant values are concentric ellipsoids, centred around the above location. Shape and location of these ellipsoids change in time so that the spatial extension of the function increases, and they move along with roughly the average velocity.

If Brownian motion is disregarded the particle motion becomes entirely deterministic. The trajectory of a particle, which was at time 0 at $\boldsymbol{r}_o$, is then given by the last equation.

This result can be obtained by letting $D \to 0$ in equation (4.10). This equation shows that for small D the value of the exponential is negligibly small unless the numerator in the exponential is close to zero. This is obviously the case if the vector $\boldsymbol{r}$ in equation (4.10) is replaced by the right hand side of equation (4.11).

4.4 The Quantitative Description of Light

Light is an electromagnetic wave which, in the framework of Maxwell's theory, is fully described by the space and time dependent fields

- $\boldsymbol{E}(\boldsymbol{r}, t)$: electric field
- $\boldsymbol{H}(\boldsymbol{r}, t)$: magnetic field

Both fields cannot be arbitrary but must be solutions of Maxwell's differential equations.

The interaction of visible light with matter (for instance with the retina of the human eye or a light detector) is, - apart from very few exceptions -, caused by the electric field only. For that reason we restrict our description of light to the electric field.

The most simple and at the same time most regular light wave is the so called plane wave. The spatial and temporal behaviour of the electric field of such a wave is given by

$$\boldsymbol{E}(\boldsymbol{r}, t) = \boldsymbol{e}E_o\cos(\boldsymbol{kr} - \omega t - \phi_0) \tag{4.12}$$

Here $\boldsymbol{e}$ is the so called polarization vector of the wave. It is dimensionless and has the length 1. It denotes the direction in which the electric field oscillates. E_o is the maximum amplitude of the wave, whereas $E_o\cos(\boldsymbol{kr} - \omega t - \phi_0)$ denotes the amplitude of the wave at time t and location $\boldsymbol{r}$. The wave vector $\boldsymbol{k}$ of light waves in isotropic media (which exhibit no dependence of optical properties on the direction in space) is always orthogonal to the polarization vector $\boldsymbol{e}$. On one hand, it denotes the direction in which the wave propagates. Everywhere on an arbitrary plane perpendicular to $\boldsymbol{k}$, the amplitude of the wave is the same for a given time. That is the reason why these waves are called plane waves. On the other hand, the length $|\boldsymbol{k}|$ of this vector is connected to the wavelength λ of the wave by the relation

$$|\boldsymbol{k}| = \frac{2\pi}{\lambda} \tag{4.13}$$

ω is the circular frequency of the wave, it is larger than the frequency ν of the wave by the of factor 2π:

$$\omega = 2\pi\nu \tag{4.14}$$

ϕ_o is the initial phase of the wave. It depends on the choice of the origin of the co-ordinate system and also on the origin of the time axis. By a special choice of one of these quantities one can always make that phase disappear. The argument of the cosine function in equation (4.12) is also called phase of the wave. Unfortunately, the two notions 'phase' and 'initial phase' are usually not clearly discriminated, the same is true for the notions 'amplitude' and 'maximum amplitude'.

Wavelength and frequency of a wave are connected with the so called phase velocity v_ϕ of the wave

$$\lambda\nu = \frac{\omega}{|\boldsymbol{k}|} = v_\phi = \frac{c}{n_M}. \tag{4.15}$$

c is the famous vacuum velocity of light and n_M the refractive index of the medium in which the wave travels. Note that a wave passing from one medium to another keeps its frequency unchanged, so generally its wavelength as well as its wave vector change on this transition.

The mathematical description of light is extremely simplified if the electric field is written in complex form according to

$$\boldsymbol{E}^c(\boldsymbol{r},t) = \boldsymbol{e}E_o \mathrm{e}^{i(\boldsymbol{kr}-\omega t-\phi_o)} \tag{4.16}$$

The physical electric field is simply the real part of $\boldsymbol{E}^c$. The product of the two factors E_o and $e^{-i\phi_o}$ can be considered as the complex amplitude E_0^c of the wave:

$$\boldsymbol{E}^c(\boldsymbol{r},t) = \boldsymbol{e}E_o^c \mathrm{e}^{i(\boldsymbol{kr}-\omega t)} \tag{4.17}$$

A light wave described by the equations (4.12), (4.16), or (4.17) is obviously a rather theoretical abstraction of any real light wave: The equations describe electric fields that are present everywhere in space and oscillate for an infinite time with the same frequency.

A real wave that comes closest to the above model is the wave of a focused laser beam in a finite region around the focus. In that region, the areas of constant phase are plane but the amplitude of the electric field decreases with increasing distance from the axis of the beam. For the description of our experiments, the assumption of the ideal time dependence of the above equations is admissible. It was shown in [12] that the electric field of this portion of the light wave can be represented with sufficient accuracy in the following way:

$$\boldsymbol{E}^c(\boldsymbol{r},t) = \boldsymbol{e}(\boldsymbol{r})E_o(\boldsymbol{r})\mathrm{e}^{i(\boldsymbol{kr}-\omega t-\phi(\boldsymbol{r})-\psi(t))}. \tag{4.18}$$

The amplitude (meaning 'maximum amplitude' from now on unless otherwise stated) of this wave is independent of time, but varies in space. For laser beams moderately focused the spatial variation of the amplitude is not very pronounced on the scale of the wavelength. However, the factor $\mathrm{e}^{i\boldsymbol{kr}}$ varies rapidly on such a scale: If $\boldsymbol{r}$ is increased (or decreased) by an amount of half a wavelength in the direction of $\boldsymbol{k}$, $\mathrm{e}^{i\boldsymbol{kr}}$ changes sign.

The quantity $\phi(\boldsymbol{r})$ also varies rather slowly with $\boldsymbol{r}$. Only in the case of a very strongly focused laser beam, the change of $\phi(\boldsymbol{r})$ becomes noticeable in the region of the geometrical focus.

$\psi(t)$ is the randomly varying phase of the laser light. This quantity is responsible for the finite correlation time τ_L of laser light. Due to the wave velocity c/n_M of the light, the coherence length ρ_L is connected with this coherence time according to $\rho_L = c/(n_M \tau_L)$.

This finite coherence length describes quantitatively the fact that light from different locations along the beam direction does not interfere if the distance of these locations is larger than ρ_L. The experimental methods described in this book only require coherence lengths of the order of the diameter of the scattering volume, which is always smaller than a millimetre. Ordinary lasers meet this condition in any case.

In regions where the wave is essentially plane, the two quantities $\phi(\boldsymbol{r})$ and $\boldsymbol{e}(\boldsymbol{r})$ can be considered constant and one can write the electric field of the wave approximately as

$$\boldsymbol{E}^c(\boldsymbol{r}, t) = \boldsymbol{e} E_o^c(\boldsymbol{r}) e^{i\boldsymbol{kr} - i\omega t + \psi(t)} \tag{4.19}$$

Thus, the region around the focus of a focused laser beam represents to a good approximation a plane wave with a spatially dependent amplitude. A closer look at properties of a laser beam will be given in Sect. 5.1.

A second type of light waves, which play an important role in the quantitative description of light scattering phenomena, are spherical waves. The phase of such a wave is given by $|\boldsymbol{k}||\boldsymbol{r} - \boldsymbol{r}_o| - \omega t$, where $\boldsymbol{r}_o$ is a constant vector denoting the centre of the spherical wave. Obviously, the phase only depends on the distance of the point $\boldsymbol{r}$ from the centre $\boldsymbol{r}_o$. If the maximum amplitude $E_o^c(\boldsymbol{r})$ does only depend on $|\boldsymbol{r}|$, i.e. if it is independent of the direction, one has an ideal spherical wave with the same amplitudes everywhere on spheres around the wave centre. Light waves, however, can only be spherical some distance away from the wave centre. Furthermore, the amplitude generally depends on the direction and the polarization vector of such an approximately spherical wave cannot be constant, but must always be perpendicular to the line connecting the wave centre and the point considered. Spherical waves in that sense are produced, for instance, by light scattering of laser light at spherical particles.

Complicated light waves, as they are for instance produced by the scattering of laser light at an ensemble of particles randomly distributed in space, can always be viewed as a superposition of plane waves or spherical waves. Superposition means that the electric fields of the composite wave is the sum of the electric fields of the component waves. It is then possible to speak either of a 'complicated wave' or a 'complicated superposition' of simple waves.

Any light detection process is based on the absorption of light and the transformation of light energy to some other energy form, which is then usually used to produce an electric signal. There is no way to measure the electric field of a light wave directly. In the experiments we are going to describe here, one is interested in the temporal behaviour of the scattered light, which means that one would like to measure the energy per time interval that hits the detector. The quantity, which is usually used to contain this information, is the light intensity, which represents the energy of a light wave per area and time. This quantity is connected with the electric field of the wave by the simple relation:

$$I(\boldsymbol{r}, t) = \frac{\epsilon_o c}{2} \boldsymbol{E}(\boldsymbol{r}, t) \cdot \boldsymbol{E}^*(\boldsymbol{r}, t). \tag{4.20}$$

In this equation ϵ_o is the electric field constant. The asterisk denotes the complex conjugate of the respective quantity. We will henceforth call the quantity $\frac{2}{\epsilon_o c} I(\boldsymbol{r}, t)$ light intensity. Even though this is not strictly correct, this notion is widely used since it renders formulae a bit more simple.

A further simplification is to dispense with the superscript ^{c} of the symbol for the electric field, as we did already in the last equation. From now on we will always understand the electric field of a light wave as a complex quantity without explicitly denoting it by this superscript.

It is worth while stressing that for the electric field of light waves the superposition principle is valid, whereas for the intensities it is not. If, for instance, a screen is illuminated by two or more light waves the intensity at a certain point is not the sum of the intensities which produce the different waves at that point. Rather, the electric fields of the light waves at that point sum up to the resulting electric field, and the intensity is then just the absolute square of this resulting field according to equation (4.20). So it may well happen that two light waves produce at some point zero intensity, while the intensities of both composing light waves are not equal zero. This effect is known as interference and the methods we are going to describe are based on this phenomenon.

4.5 Light Scattering by Particles

We are now going to consider the properties of light scattered by an ensemble of small particles in a flow. The flow should be illuminated by a plane wave, and the particles are assumed to be spherical, and have the same size and optical properties.

Let us first consider the wave originating from scattering of a linearly polarized light wave at an isolated particle. As we mention above (c.f. Sect. 4.4) the scattered wave can, far away from the particle, be considered as a spherical wave. Since the amplitude of the scattered wave depends on the observation direction with respect to the polarization direction, one usually considers two main scattering geometries. They are defined by the polarization vector of the illuminating light wave and the scattering plane. This is the plane formed by the direction of the illuminating light wave and the observation direction.

The two mentioned scattering geometries are:

- σ-scattering. The polarization vector is perpendicular to the scattering plane.
- π-scattering. The polarization vector lies in the scattering plane.

In light scattering experiments, the scattered light is usually observed in a particular direction as a function of time. For a theoretical description it is then useful to view the scattered wave as composition of different plane waves. Clearly, only one of these plane wave components travels in the required direction, all other components are entirely irrelevant. The important component is marked by the direction of its wave vector, which must be parallel to the observation direction. The situation for the two discussed scattering geometries is depicted in Fig. 4.1.

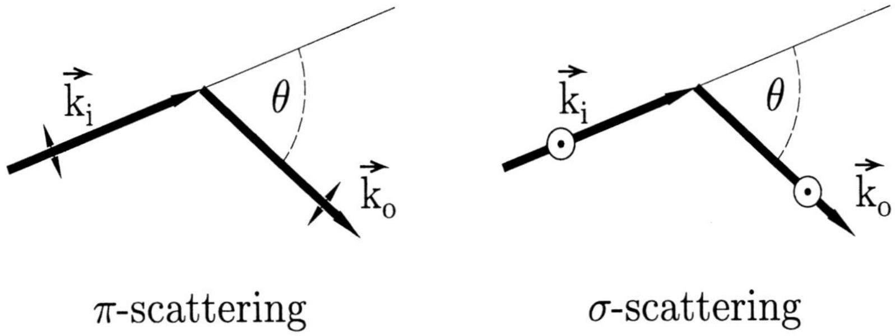

Fig. 4.1. The two main scattering geometries. $\boldsymbol{k}_i$ is the wave vector of the illuminating light wave, $\boldsymbol{k}_o$ is that of the observed light component. The symbols attached to these vectors denote the respective polarizations of the scattered waves. Θ is the scattering angle

The amplitude of the scattered light is given by a quantity that depends on the ratio of the refractive indices n_F (refractive index of the fluid) to n_P (refractive index of the particle), the ratio of particle diameter d to wavelength of the light λ, and the distance L of the observation point from the particle. The situation is particularly simple for particles that are small compared to the wavelength. In case of σ-scattering, the scattered light amplitude is then independent of the scattering angle, however, for π-scattering a significant dependence does exist:

$$\pi\text{-scattering: } E_S = E_o Q \frac{\left(\frac{d}{\lambda}\right)^2 \left(\left(\frac{n_P}{n_F}\right)^2 - 1\right)}{L} \cdot \cos(\theta) \tag{4.21}$$

$$\sigma\text{-scattering: } E_S = E_o Q \frac{\left(\frac{d}{\lambda}\right)^2 \left(\left(\frac{n_P}{n_F}\right)^2 - 1\right)}{L}$$

Here E_o represents the amplitude of the illuminating light wave at the particle. The constant Q depends in a complex way on properties of the particle, it can be calculated numerically (see e.g.[13]). This factor is not important for our considerations.

The factor of the quantity E_o in equation (4.21) will be called the scattering factor of the particle, σ, and we then obtain the simple equation:

$$E_S = \sigma E_o \tag{4.22}$$

This equation is also valid for particles not small compared to the wavelength, even though the scattering factor in that case becomes a complicated function of the scattering angle and the two ratios defined above (see discussion prior to equation (4.21).

For non-spherical particles one has also a dependence of the scattering factor on the orientation of the particle with respect to the scattering plane and polarization vector. Furthermore, such particles will generally produce elliptically polarized light. Particles of this type should therefore not be used for the measurement techniques described here.

The more general case of scattering at many particles does now present no conceptual difficulty: The total light wave scattered by the ensemble of particles is just the sum of the light waves scattered by single particles. However, this can be only approximately true, because the light wave scattered at one particle will be scattered again by other particles and so on, giving rise to the so called multiple scattering. Furthermore, the primary light wave becomes attenuated on its way through the scattering medium by the scattering processes themselves. The simple picture can therefore only be adopted if the scattering efficiency of one particle is very small and there are not too many particles within the scattering volume, so that the attenuation of the illuminating light beam in the scattering volume is negligible. In this chapter, we will exclusively deal with single scattering, multiple scattering will be discussed later on (Sect. 5.3).

Having made these considerations, we are now able to treat quantitatively the basic light scattering experiment shown in Fig. 4.2.

A plane light wave travels through a fluid containing light scattering particles. A fraction of the scattered light falls onto the lens L. In the focal plane of this lens an observation screen is placed. This set-up achieves that the light component which travels in a definite direction hits the screen at exactly one point. In other words: The light intensity at a particular point on the screen is determined by exactly one plane wave component of the scattered light field. We will denote the wave vector of the component responsible for the intensity at the point P by $\boldsymbol{k}_o$. This component of the scattered wave can itself be considered as the sum of the corresponding plane wave components of the scattered waves of every single particle. This is indicated in the figure by the dashed 'light rays' originating at the scattering particles and ending up at the point P. 'Rays' in other directions will be focused to other points on that screen.

The auxiliary plane E is constructed perpendicular to the optical axis of the focusing lens. With an ideal lens all optical light paths for arbitrary parallel rays beginning at the plane E and ending up at their common focus point on the observation plane have the same length. This means that phase differences on the plane E and at point P are identical. It is therefore possible to obtain the intensity at a point on the observation screen by calculating the intensity of the corresponding plane wave components on the plane E.

We are now going to calculate the electric field of the scattered light wave at an arbitrary point on the observation screen with the approximations just discussed. Let us write the electric field of the illuminating plane light wave in the plane A (which is constructed to be a plane of constant phase) in the form

$$\boldsymbol{E}_A(t) = \boldsymbol{e}_e E_o \exp\left(-i(\omega t + \phi_o)\right) \tag{4.23}$$

where ϕ_o is the phase at time zero. It is not difficult to calculate the phase of this wave at an arbitrary particle which is located at the point $\boldsymbol{r}$. Clearly, the light has to travel a distance L_1, which is simply the projection of the vector $\boldsymbol{r}$ in the direction of the illuminating beam. Since the direction of this beam is given by $\boldsymbol{k}_i$ we immediately obtain:

$$L_1 = \boldsymbol{r} \cdot \frac{\boldsymbol{k}_i}{|\boldsymbol{k}_i|} \tag{4.24}$$

The phase difference of the illuminating light beam between the plane A and the location $\boldsymbol{r}$ is then obtained by dividing the path length by λ and multiplying it by 2π:

$$\Delta_1 \phi = \boldsymbol{r} \cdot \frac{\boldsymbol{k}_i}{|\boldsymbol{k}_i|} \frac{2\pi}{\lambda} = \boldsymbol{r} \cdot \boldsymbol{k}_i \tag{4.25}$$

The illuminating light wave at the particle has therefore the form

$$\boldsymbol{E}(\boldsymbol{r}, t) = \boldsymbol{e}_e E_o \exp\left(-i(\omega t + \phi_o + \boldsymbol{r} \cdot \boldsymbol{k}_i)\right) \tag{4.26}$$

The particle hit by the illuminating wave produces a spherical wave. We are interested in that component of this wave which travels in direction $\boldsymbol{k}_o$. It has an amplitude proportional to the scattering factor σ and to the electric field of the illuminating wave at its location. We will now calculate the phase of this component in the plane E. The path the light has to travel from its origin to E will be denoted by L_2. An analogous calculation, as done above, gives the phase difference of this wave between $\boldsymbol{r}$ and the plane E:

$$\Delta_2 \phi = -\boldsymbol{r} \cdot \frac{\boldsymbol{k}_o}{|\boldsymbol{k}_o|} \frac{2\pi}{\lambda} = -\boldsymbol{r} \cdot \boldsymbol{k}_o \tag{4.27}$$

(The minus sign results from the fact that the light path is the projection of $\boldsymbol{r}$ in the direction $-\boldsymbol{k}_o$!).

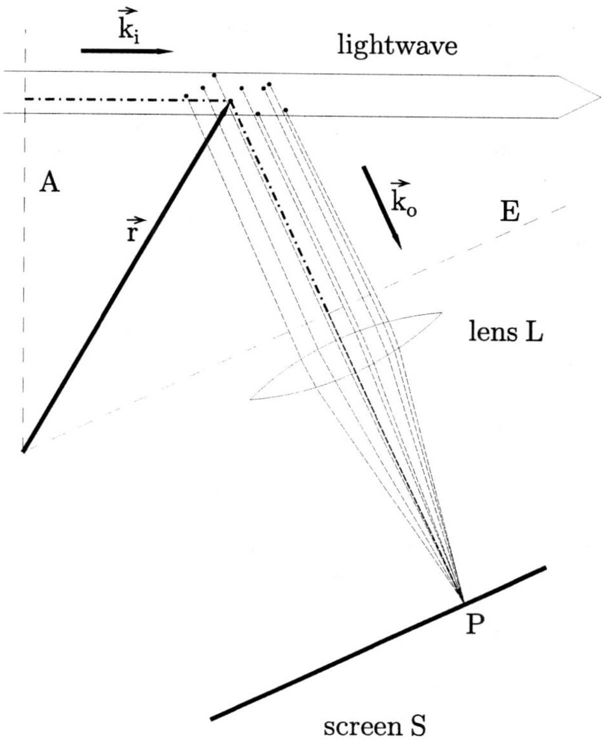

Fig. 4.2. Scheme of a light scattering experiment. The illuminating light wave is a plane wave characterized by the wave vector $\boldsymbol{k}_i$. All plane wave components of the scattered light with the wave vector $\boldsymbol{k}_o$ are focused by the lens L onto *one* point P on the observation screen, because the screen plane coincides with the focal plane of the lens. For further explanations see text.

The overall phase difference is therefore

$$\Delta\phi = \Delta_1\phi + \Delta_2\phi = \boldsymbol{r} \cdot (\boldsymbol{k}_i - \boldsymbol{k}_o) \tag{4.28}$$

As we already noted, all these waves get the same phase differences ϕ_D on their way from the plane E to the point P.

The amplitude of the scattered wave at point P for many particles is then finally given by

$$E_P(t) = \sigma \sum_{j=1}^{N} E_o \exp\left(-i\left(\omega t + \boldsymbol{r}_j(t) \cdot (\boldsymbol{k}_i - \boldsymbol{k}_o)\right)\right). \tag{4.29}$$

$\boldsymbol{r}_j(t)$ indicates the location of the j-th particle, which is generally time dependent. The phase factor $\exp(i\phi_o + i\phi_D)$ was omitted in this equation because it is of no physical relevance.

To end up with equation (4.29) we have tacitly assumed that the amplitudes of the spherical waves of all particles are identical. This, however, is not true for two reasons. Firstly, the illuminating light beam usually has a lateral intensity profile, which means that particles near the axis of the beam 'see' the most intense light field, whereas particles farther away from this axis are illuminated less. Secondly, a greater proportion of the scattered light wave of particles near the optical axis of the lens L is transmitted to the screen than is the case for particles farther away from this axis. It is possible to describe this fact quantitatively by a so called spatial aperture function $G(\boldsymbol{r})$. It is a measure of the relative amplitude of the scattered light wave on the screen, if only *one* scattering particle is present at location $\boldsymbol{r}$. We will normalize $G(\boldsymbol{r})$ in such a way that its maximum value is 1.

In many qualitative considerations it is useful to introduce the notion of the scattering volume. This is that region in space where this function is larger than a certain fraction of its maximum value, for instance e^{-1}.

In this more refined model we obtain for the amplitude of the scattered light wave at the point P on the screen:

$$E_P(t) = E(\boldsymbol{q}, t) = \sigma \sum_{j=1}^{N} E_o G(\boldsymbol{r}_j(t)) \exp(-i\omega t + i\boldsymbol{r}_j(t) \cdot \boldsymbol{q}). \tag{4.30}$$

In the experiment shown schematically in Fig. 4.2, the location of a point P in the screen plane is determined by the wave vector $\boldsymbol{k}_o$. In quantitative descriptions of scattering experiments, this vector occurs always in the form of the difference of the two important wave vectors $\boldsymbol{k}_i - \boldsymbol{k}_o$. For that reason this difference is called the scattering vector $\boldsymbol{q}$:

$$\boldsymbol{q} = \boldsymbol{k}_i - \boldsymbol{k}_o \tag{4.31}$$

For a fixed $\boldsymbol{k}_i$, i.e. a fixed, plane illuminating light wave, it is therefore always possible to indicate a point P by the corresponding scattering vector $\boldsymbol{q}$. This is the reason why we introduced the notation $E(\boldsymbol{q}, t)$ in equation (4.30) for the field amplitude at the point P.

It is useful for the ease of following discussions to draw the vectors $\boldsymbol{q}$, $\boldsymbol{k}_i$, and $\boldsymbol{k}_o$ in the way as it is shown in Fig. 4.3. In addition to the three vectors, the figure also shows the screen and that point on the screen which corresponds to $\boldsymbol{k}_o$.

Due to equation (4.13), the length of the wave vectors is constant, it is only determined by the wavelength of the laser light. The change of the wavelength caused by a motion of the scattering particle is negligibly small. However, the length of the scattering vector changes drastically with the scattering angle. It is obvious from Fig. 4.3 that by varying the scattering

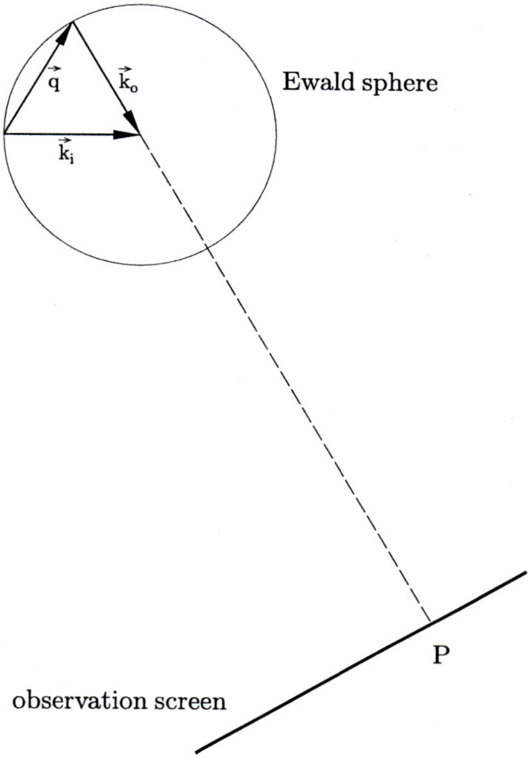

Fig. 4.3. Schematic representation of the most important elements for a theoretical description of the scattering experiment shown in Fig. 4.2. The geometrical connections are shown between the wave vectors and the scattering vector on one hand, and the wave vector $\boldsymbol{k}_o$ and the corresponding point on the observation screen on the other hand.

angle starting and end points of all possible scattering vectors lie on a sphere, called after a German physicist Ewald sphere.

In this book we always consider only small fractions of the Ewald sphere which can then be approximated by a plane. Within these limits, the observation screen directly corresponds to a small part of the Ewald sphere.

The considerations of the light scattering by a ensemble of particles led us to a sum over the particles (equation (4.30)). It is worthwhile reformulating this result to obtain a relation which can be applied much easier to general light scattering fluids. To do so, we introduce a function $K(\boldsymbol{r}, t)dV$ describing which fraction of the illuminating light is scattered at time t in a small volume dV around the location $\boldsymbol{r}$. This function may be called the optical inhomogeneity function. In the case of the light scattering particles considered so far, this function $K(\boldsymbol{r}, t)$ is equal to zero everywhere except at the location of the particles.

The adoption of this function bears the big advantage that by introducing it in equation (4.30) the summation of the particles can be replaced by a spatial integration:

$$E(\boldsymbol{q},t) = \sigma^c \int\limits_V E_o G(\boldsymbol{r}) K(\boldsymbol{r},t) \exp(-i\omega t + i\boldsymbol{r}\cdot\boldsymbol{q}) dV. \tag{4.32}$$

Finally, let us consider the ideal case where the scattering volume is infinitely large and the spatial aperture function is constant. We then have:

$$E(\boldsymbol{q},t) = \sigma^c \cdot L \cdot E_o \cdot \exp(-i\omega t) \int\limits_V K(\boldsymbol{r},t) \exp(i\boldsymbol{r}\cdot\boldsymbol{q}) dV. \tag{4.33}$$

The integral is clearly just the Fourier transform $\tilde{K}(\boldsymbol{q},t)$ of the optical inhomogeneity:

$$\tilde{K}(\boldsymbol{q},t) = \int K(\boldsymbol{r},t) e^{i\boldsymbol{q}\boldsymbol{r}} dV. \tag{4.34}$$

This result can now be expressed in the following way: The amplitude of light scattered in the direction $\boldsymbol{k}_o$ is only determined by the Fourier component of the optical inhomogeneity belonging to the vector $\boldsymbol{k}_i - \boldsymbol{k}_o = \boldsymbol{q}$. For scattering volumes with a finite size this result is, of course, only approximately true, but this approximation is usually sufficient in cases where the extensions of the scattering volume are large compared with the wavelength.

The mathematical form of an optical inhomogeneity with a specific q-vector is given by the expression $K(\boldsymbol{r},t) = K_{\boldsymbol{q}}(t)\exp(i\boldsymbol{q}\cdot\boldsymbol{r})$. A physical inhomogeneity $K(\boldsymbol{r},t)$, however, can never be of this type because it must always be real and can never be described by a complex quantity. A physically possible form can be the following:

$$K(\boldsymbol{r},t) = K_o(t)\left(\exp(i\boldsymbol{q}\cdot\boldsymbol{r}) + \exp(-i\boldsymbol{q}\cdot\boldsymbol{r})\right) = 2K_o(t)\cos(\boldsymbol{q}\cdot\boldsymbol{r}) \tag{4.35}$$

where $K_o(t)$ is assumed to be real.

If equation (4.35) describes the spatial variation of the refractive index, the optical inhomogeneity can be called a 'frozen' refractive index wave. The loci with maximal refractive index are planes perpendicular to $\boldsymbol{q}$, the wave vector of this wave. The distance of neighbouring planes is $2\pi/|\boldsymbol{q}|$, according to equation (4.13).

If a medium was illuminated by that type of refractive index wave, it would only be possible to observe scattered waves in two directions. For an illuminating light wave with wave vector $\boldsymbol{k}_i$, these two directions are $\boldsymbol{k}_i + \boldsymbol{q}$ and $\boldsymbol{k}_i - \boldsymbol{q}$. However, scattering in one of these directions only occurs if one of the two vectors

$$\boldsymbol{k}_i = \boldsymbol{k}_i \pm \boldsymbol{q} \tag{4.36}$$

can actually be the wave vector of a scattered wave. This is only then the case when $\boldsymbol{k}_i$ has the same length as $\boldsymbol{k}_i$:

$$|\boldsymbol{k}_i| = |\boldsymbol{k}_i| = \frac{2\pi}{\lambda}. \tag{4.37}$$

One can easily see that for a finite $\boldsymbol{q}$ this last condition can only be fulfilled for one sign in equation (4.36). For an arbitrary orientation of the illuminating light beam with respect to the refractive index wave, one will therefore observe no scattered light whatsoever, which seems rather strange, because there is an optical inhomogeneity in the medium. One can argue that in this general case the spherical waves originating everywhere within the medium interfere destructively in any direction, except in the forward direction.

Adopting the general optical terminology, the above would not be called light scattering, but diffraction of light by a spatial grating.

4.6 The q-Volume of Coherence

Let us consider the scattered light which is obtained if an irregular light scattering object, as for instance white paper, is illuminated by a laser beam. The scattered light is observed on a screen which is far away - about a meter - from the bright laser spot on the object; the size of this illuminated spot should be roughly a millimetre. One will then observe the so called speckle pattern on the screen, a random pattern which is caused by interference. Examples of such speckle patterns are shown in Fig. 4.4.

The patterns shown there are synthetic patterns. They were produced on the basis of equation (4.30). The locations of 90 particles were obtained by a random number generator, where a three dimensional Gaussian function served as aperture function $G(\boldsymbol{r})$. The Gaussian for the picture on the left hand side was spherically symmetric, for the other picture we chose a horizontal extension of the Gaussian that was half that of the vertical extension.

Although the pattern varies in an erratic way on the screen, one will notice that the bright spots can be characterized by their size. This can easily be demonstrated if the size of the illuminating laser spot is varied: By focusing it down, one observes a marked increase in the size of the speckles, whereas an increase of the spot shrinks the speckle size.

This type of pattern of course also arises if a fluid containing light scattering particles is illuminated by a laser beam. But the particles will undergo Brownian motion and therefore the pattern will change in time. If this temporal change is fast, the human eye will not be able to resolve the time dependence and one will observe only an average intensity. In the forward direction, however, the pattern motion will be slow enough, and the time dependent speckle pattern can be observed well.

The simplest way to quantify the size of the speckles is based on the spatial correlation function of the speckle pattern. One can either consider the amplitude correlation function

$$C_E(\boldsymbol{r}_1, \boldsymbol{r}_2, 0) = \langle E(\boldsymbol{r}_1, t) \cdot E^*(\boldsymbol{r}_2, t) \rangle \tag{4.38}$$

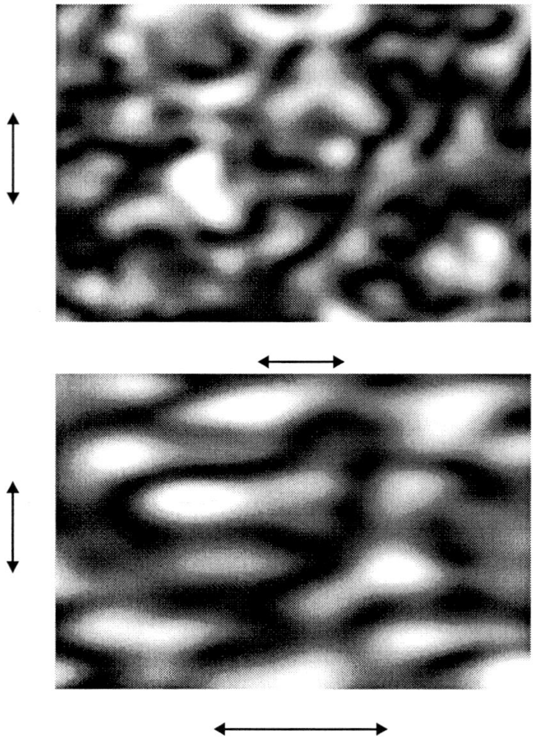

Fig. 4.4. Examples of speckle patterns . The arrows indicate the diameters of the coherence areas in the corresponding directions.

or the intensity correlation function

$$C_I(\boldsymbol{r}_1, \boldsymbol{r}_2, 0) = \langle I(\boldsymbol{r}_1, t) \cdot I(\boldsymbol{r}_2, t) \rangle \,. \tag{4.39}$$

(The vectors $\boldsymbol{r}_1$ and $\boldsymbol{r}_2$ are here two-dimensional vectors on the observation screen.) The notation indicates that we do not require homogeneity of the pattern, but stationarity is necessary.

There is a simple relation between these two correlation function for the big class of Gaussian light fields, i.e. those where the probability density $P(E)$ for the amplitude of the light field is a Gaussian. In that case one has:

$$\begin{aligned} C_I(\boldsymbol{r}_1, \boldsymbol{r}_2, \tau) &= C_E(\boldsymbol{r}_1, \boldsymbol{r}_1, 0) \cdot C_E(\boldsymbol{r}_2, \boldsymbol{r}_2, \tau) + |C_E(\boldsymbol{r}_1, \boldsymbol{r}_2, \tau)|^2 \quad (4.40) \\ &= \langle I(\boldsymbol{r}_1, t) \rangle \langle I(\boldsymbol{r}_2, t + \tau) \rangle + |C_E(\boldsymbol{r}_1, \boldsymbol{r}_2, \tau)|^2 \end{aligned}$$

This relationship is known as 'Siegert relation'. It is, for instance, always true, if the scattered light results from scattering by a body with a random optical inhomogeneity, either on the surface or within it. The *exact* calculation of one of the two correlation functions for randomly distributed scatters, is unfortunately so sophisticated that we will not show it here. Instead, we will show for a special case how the general behaviour of the speckle size depends on certain parameters of the scattering experiment.

Let us then consider the following case: The point-like scattering particles shall be distributed in an entirely random way, i.e. the probability to find the n-th particle somewhere shall be independent of where the other particles are (since particles with small distances always have some interaction, this assumption will only be justified if the particle concentration is sufficiently low). We furthermore assume a Gaussian shape for the spatial aperture function $G(\boldsymbol{r})$. This choice is motivated by the fact that in this case the necessary integration can be carried out analytically. However, a more general treatment shows that the correlation function does not depend sensitively on the exact form of the aperture function. That feature of the aperture function, that in the most marked manner determines the correlation function, is the extension of it in the different directions in space.

We will now give a rough derivation of the amplitude correlation function. Let us recall that we may denote a particular point on the observation screen by the vector $\boldsymbol{q}$. The spatial amplitude correlation function can then be written as:

$$C_E(\boldsymbol{q}_1, \boldsymbol{q}_2, 0) = \langle \boldsymbol{E}(\boldsymbol{q}_1, t) \cdot \boldsymbol{E}^*(\boldsymbol{q}_2, t) \rangle \tag{4.41}$$

Introduction of the electric fields from equation (4.32) yields

$$C_E(\boldsymbol{q}_1, \boldsymbol{q}_2, 0) = \Big\langle AE_o \exp(-i\omega t) \int_V G(\boldsymbol{r}_1) K(\boldsymbol{r}_1, t) \exp(i\boldsymbol{r}_1 \cdot \boldsymbol{q}_1) dV_1 \cdot \tag{4.42}$$
$$A^* E_o \exp(i\omega t) \int_V G(\boldsymbol{r}_2) K(\boldsymbol{r}_2, t) \exp(-i\boldsymbol{r}_2 \cdot \boldsymbol{q}_2) dV_2 \Big\rangle .$$

The function $K(\boldsymbol{r}_1, t)$ is evidently proportional to the particle density $\rho(\boldsymbol{r}_1, t)$, so we may write

$$K(\boldsymbol{r}_1, t) = K_o \rho(\boldsymbol{r}_1, t) \tag{4.43}$$

The only stochastic quantity, which appears in equation 4.42 and which must be averaged, is the product $K(\boldsymbol{r}_1, t) \cdot K(\boldsymbol{r}_2, t)$. By using equation 4.43 we can then write

$$C_E(\boldsymbol{q}_1, \boldsymbol{q}_2, 0) = |A|^2 |E_o|^2 K_o^2 \int_V \int_V G(\boldsymbol{r}_1) G(\boldsymbol{r}_2) \langle \rho(\boldsymbol{r}_1, t) \rho(\boldsymbol{r}_2, t) \rangle \tag{4.44}$$
$$\exp(i\boldsymbol{r}_1 \cdot \boldsymbol{q}_1 - i\boldsymbol{r}_2 \cdot \boldsymbol{q}_2) dV_1 \cdot dV_2 .$$

In analogy to equation (3.29), one can express the ensemble average in this equation by an integration. However, we are not going to write down the exact form of the joint probability function and then actually calculate the integral, since this procedure would be beyond the scope of this book. Let us instead make plausibility assumptions on the important properties of the probability $P^2(\rho, \rho'; \boldsymbol{r}_1, \boldsymbol{r}_2, t, t)dV1_1 dV_2$ for the case of interest here. (See also the derivation of equation 4.32). At first we express the joint probability function with the help of a conditional probability:

$$P^2(\rho, \rho'; \boldsymbol{r}_1, \boldsymbol{r}_2, t, t)dV_1 dV_2 = P_c(\rho; \boldsymbol{r}_1, t|\rho'; \boldsymbol{r}_2, t)dV_1 \cdot P^1(\rho'; \boldsymbol{r}_2, t)dV_2 \quad (4.45)$$

The second factor on the right hand side is a constant, since the probability to find a particle somewhere is independent of space and time. For the other factor, two contributions can be discriminated: The first one arises if we put $\boldsymbol{r}_1 = \boldsymbol{r}_2$. In that case $P_c \cdot dV_1$ gives the probability to find a particle in the volume dV_1 around the location $\boldsymbol{r}_1$, provided it is located at $\boldsymbol{r}_1$. This probability is 1.

If $\boldsymbol{r}_1 \neq \boldsymbol{r}_2$ there is, of course, a certain probability to find another particle in the volume dV_1 at $\boldsymbol{r}_1$ which is different from that at $\boldsymbol{r}_2$. This probability shall be, however, independent of $\boldsymbol{r}_1$, as we stated above. This probability is therefore independent of $\boldsymbol{r}_1$, which means it is constant.

The average $\langle \rho(\boldsymbol{r}_1, t)\, \rho(\boldsymbol{r}_2, t) \rangle$ is then composed of two contributions. The first one is different from zero only for $\boldsymbol{r}_1 = \boldsymbol{r}_2$, and the second one is constant. Thus, the correlation function can be written as the sum of two parts, which we will call C_{E1} and C_{E2}.

C_{E1} is obtained if we put $\boldsymbol{r}_1 = \boldsymbol{r}_2$ on the right hand side of equation (4.44) and integrate over dV_1:

$$C_{E1}(\boldsymbol{q}_1, \boldsymbol{q}_2, 0) = K_o^2 A^2 \int_V E_o^2 G^2(\boldsymbol{r}_1) \exp(i\boldsymbol{r}_1 \cdot (\boldsymbol{q}_1 - \boldsymbol{q}_2)) dV_1. \quad (4.46)$$

One observes that this part of the correlation function only depends on the difference $\boldsymbol{q}_2 - \boldsymbol{q}_1 = \boldsymbol{\kappa}$ of the two q-vectors. For this reason we write this correlation function simply in the form $C_{E1}(\boldsymbol{\kappa}, 0)$. The calculation of the integral in equation (4.46) requires the specification of the spatial aperture function. As already stated, we assume a spatial Gaussian which we write in the form

$$G(\boldsymbol{r}) = \exp\left(-\frac{x^2}{2d_x^2} - \frac{y^2}{2d_y^2} - \frac{z^2}{2d_z^2}\right). \quad (4.47)$$

The extensions in the three directions of an orthogonal co-ordinate system are given by the three d_i. The integration can now be carried out and yields

$$C_{E1}(\boldsymbol{\kappa}, 0) = \tilde{C} \exp\left(-\frac{1}{4}\left(\kappa_x^2 d_x^2 + \kappa_y^2 d_y^2 + \kappa_z^2 d_z^2\right)\right). \quad (4.48)$$

The factor $\tilde{C}$ contains all those constants and parameters which are not of interest for the following. We then have the result that the amplitude correlation function is a Gaussian in the variable $\boldsymbol{\kappa}$, which is a measure of the distance of two points in the observation plane.

We are not going to calculate the second part C_{E2} of the correlation function in the same way as we did the first part. The integral in equation (4.44) can be written as a product of two integrals, one over dV_1 and the other over dV_2. The integrations yield

$$C_{E2}(\boldsymbol{\kappa},0) = \tilde{C}\exp\left(-\frac{1}{2}\left(q_{1x}^2 d_x^2 + q_{1y}^2 d_y^2 + q_{1z}^2 d_z^2\right)\right)\exp\left(-\frac{1}{2}\left(q_{2x}^2 d_x^2 + q_{2y}^2 d_y^2 + q_{2z}^2 d_z^2\right)\right). \tag{4.49}$$

This part of the correlation function is essentially only different from zero for very small $\boldsymbol{q}_1$ and $\boldsymbol{q}_2$. Physically, this means that it gives a contribution to the correlation function only for forward scattering, and this contribution is of no interest for our purposes.

The correlation function C_{E1} now gives us the extent to which the scattered light at two different locations on the screen is correlated. In the case of light, the correlation function is directly connected with the notion of coherence. Coherence is quantified by the requirement that its maximum value is 1, and that this maximum always occurs if the two points under consideration are chosen to be the same. The correlation function we just calculated takes on its maximum value for $\boldsymbol{\kappa} = 0$, which means the two points on the screen are identical. Then the normalized correlation function is given by:

$$c_E(\boldsymbol{\kappa},0) = \exp\left(-\frac{1}{4}\left(\kappa_x^2 d_x^2 + \kappa_y^2 d_y^2 + \kappa_z^2 d_z^2\right)\right). \tag{4.50}$$

By use of equation (4.40) we can obtain the intensity correlation function, which we will also give in its normalized version:

$$c_I(\boldsymbol{\kappa},0) = 1 + \exp\left(-\frac{1}{2}\left(\kappa_x^2 d_x^2 + \kappa_y^2 d_y^2 + \kappa_z^2 d_z^2\right)\right). \tag{4.51}$$

It is generally useful to consider the normalized correlation function of the intensity fluctuations, which is obtained from the intensity correlation function by subtracting the constant background (which is the value of the correlation function for large distances):

$$\tilde{c}_I(\boldsymbol{\kappa},0) = \exp\left(-\frac{1}{2}\left(\kappa_x^2 d_x^2 + \kappa_y^2 d_y^2 + \kappa_z^2 d_z^2\right)\right). \tag{4.52}$$

Light on the screen at two locations $\boldsymbol{q}_1$ and $\boldsymbol{q}_2$ is now called (partially) coherent, if the normalized amplitude, i.e. the electric field correlation function for $\boldsymbol{\kappa} = \boldsymbol{q}_1 - \boldsymbol{q}_2 = -\boldsymbol{k}_{o1} + \boldsymbol{k}_{o2}$, is larger than 1/e. Otherwise light is called incoherent.

Let us now take an arbitrary point $\boldsymbol{q}_1$ on the observation screen and mark all points $\boldsymbol{q}_2$ that have the property $c_E(\boldsymbol{q}_1 - \boldsymbol{q}) = 1/e$. Obviously, these points form a closed curve which defines the so called *coherence area.*

These relationships are shown schematically in Fig. 4.5. The light at the two locations P_1 and P_2 is said to be coherent if the correlation function $c_E(\boldsymbol{\kappa})$ is bigger than 1/e. One can furthermore deduce from this drawing that with a fixed $\boldsymbol{k}_{o1}$ the starting and end points of all $\boldsymbol{\kappa}$ must lie on the Ewald sphere. Vectors $\boldsymbol{\kappa}$, which do not fulfill this condition, do not describe scattered light, but are entirely fictitious.

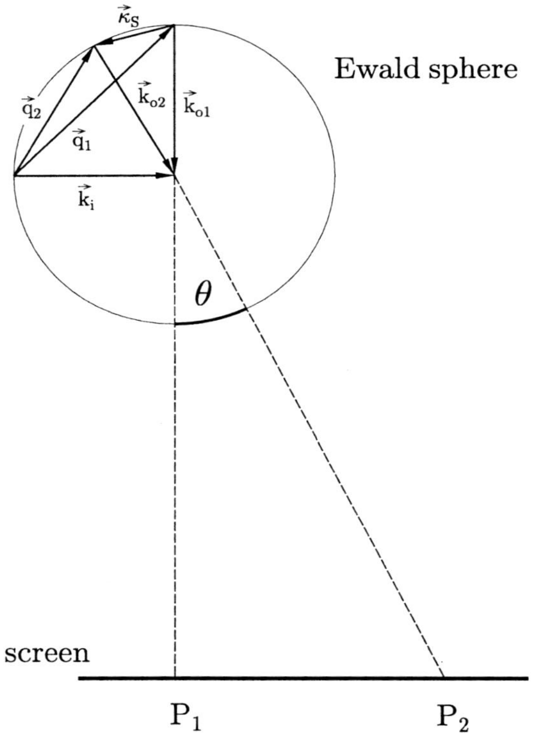

Fig. 4.5. The correlation of the scattered light at two locations on the observation screen. Further explanations in the text.

If one expresses the just mentioned relationships via the intensity correlation function $\tilde{c}_I$, one has to take into account that the intensity depends quadratically on the electric field; therefore, the value of the function $\tilde{c}_I$ is $1/\mathrm{e}^2$ at the margin of the coherence area.

The relation between the diameters of the scattering volume and the coherence area is depicted in Fig. 4.6. A light scattering surface is illuminated by an elliptic laser beam. The screen is then illuminated by the scattered

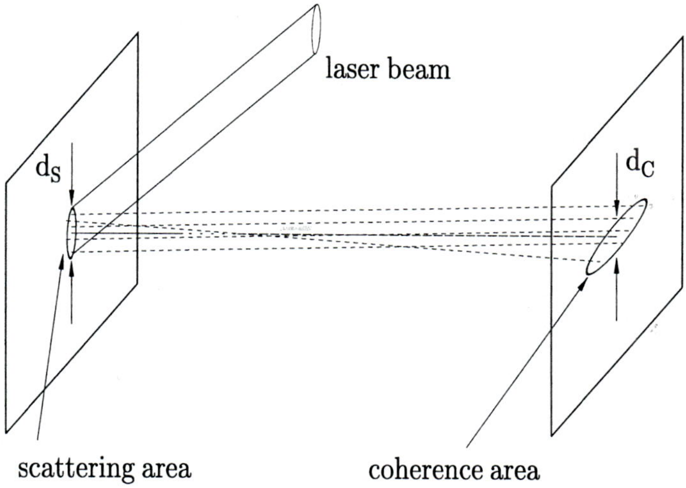

Fig. 4.6. The relation between the extensions of the scattering volume and the coherence area. One has $\mathrm{d}_C \sim \mathrm{d}_S^{-1}$.

light. On the screen, a speckle pattern is observed with a coherence area of the indicated form. Of course, the coherence area cannot be observed directly, it can only be obtained by a spatial averaging process. However, visual inspection of the speckle pattern clearly shows a stretched form of the speckles. This is shown in Fig. 4.4 on the right hand side.

We are now going to calculate the connection between the diameter of the coherence area, d_C, and the diameter of the scattering area, d_S. For that purpose, let us consider the Fig. 4.5. Let us assume that for a fixed vector $\boldsymbol{q}_1$, the vector $\boldsymbol{q}_2$ is chosen in such a way that the normalized correlation function $c_E(\boldsymbol{\kappa}_S = \boldsymbol{q}_1 - \boldsymbol{q}_2)$ assumes just the value e^{-1}, which defines the vector $\boldsymbol{\kappa}_S$. We then obtain, for reasonable scattering volumes, i.e. $d_S > 100 \cdot \lambda$, values of the angle θ which are much smaller than that shown in the figure. That means that the vector $\boldsymbol{\kappa}_S$ is, to a good approximation, parallel to the line $\overline{P_1 P_2}$. The direction of this line will be chosen to coincide with the x-axis of a co-ordinate system. Using equation (4.52), we then obtain with $\kappa_y = \kappa_z = 0$

$$\kappa_S = \frac{\sqrt{2}}{d_x}. \tag{4.53}$$

By inspection of Fig. 4.5 we furthermore deduce

$$\frac{\kappa_S}{|\boldsymbol{k}|} = \frac{\overline{P_1 P_2}}{L} = \frac{d_C}{2L}, \tag{4.54}$$

where L is the distance of the screen to the scattering volume, and $d_C/2$ is by definition half the diameter of the coherence area and at the same time the distance of those two points on the screen, which are indicated by the two scattering vectors $\boldsymbol{q}_1$ and $\boldsymbol{q}_2$. Inserting equation (4.54) into equation (4.53) gives

$$d_C = L \cdot \lambda \cdot \frac{\sqrt{2}}{\pi d_x}. \tag{4.55}$$

If the diameter d_S of the scattering volume is defined by the distance of the two points, where the aperture function dropped to the e^{-1}-th part of its value in the centre, one gets with the help of equation (4.47)

$$d_x = \frac{d_S}{2\sqrt{2}} \tag{4.56}$$

Inserting this equation into equation (4.55) leads, together with equation (4.47), to the final result:

$$d_C = \frac{4}{\pi} \frac{L \cdot \lambda}{d_S}. \tag{4.57}$$

In Fig. 4.4, the diameters d_C are indicated by arrows.

By using two illuminating light waves and observing the scattered light, which originates in the volume that is illuminated by both rays, it is possible to vary the vector $\boldsymbol{\kappa}$ not only on a surface, but three dimensionally. This is illustrated in Fig. 4.7. The wave vectors of the two illuminating rays are $\boldsymbol{k}_{i1}$ (ray 1) and $\boldsymbol{k}_{i2}$ (ray 2). The experiment should now provide for the possibility that only two components of the scattered light are observed: That of the illuminating ray 1 with the wave vector $\boldsymbol{k}_{o1}$, and the other of ray 2 with the wave vector $\boldsymbol{k}_{o2}$. There are several possibilities to achieve that. If one uses two detectors to measure the intensities at points P_1 and P_2, respectively, one could turn off the two illuminating light beams alternatively and to gate the detector signals with the same frequency, such that only the detector 1 is active when light ray 1 illuminates the sample, and vice versa. An other one would be to use lasers with different wavelengths and place optical filters in front of the detectors so that only the desired scattered light reaches the corresponding detector. If the q-vectors of these two scattering experiments are equal, then the two observed intensities are equal as well, since the property of the scattered light only depends on the scattering vector $\boldsymbol{q} = \boldsymbol{k}_i - \boldsymbol{k}_o$. Controlled variation of the directions of the illuminating light beams and of the observation directions allows for a continuous change of the vector $\boldsymbol{\kappa}$ in a certain volume in k-space.

By this method it is possible to measure the spatial correlation function. The set of all vectors $\boldsymbol{\kappa}$ with the property $\tilde{c}_I(\boldsymbol{\kappa}) = 1/e^2$ now forms a closed surface, which is the surface of a volume which we call q-volume of coherence. It is worth noting that the usual definition of the term coherence volume refers to a property of a light field, whereas the q-volume of coherence introduced here is rather a property of the light scattering object. Nevertheless, the notion coherence area which we introduced above, refers to the same property as the usual definition. It is simply given by the area which this surface of

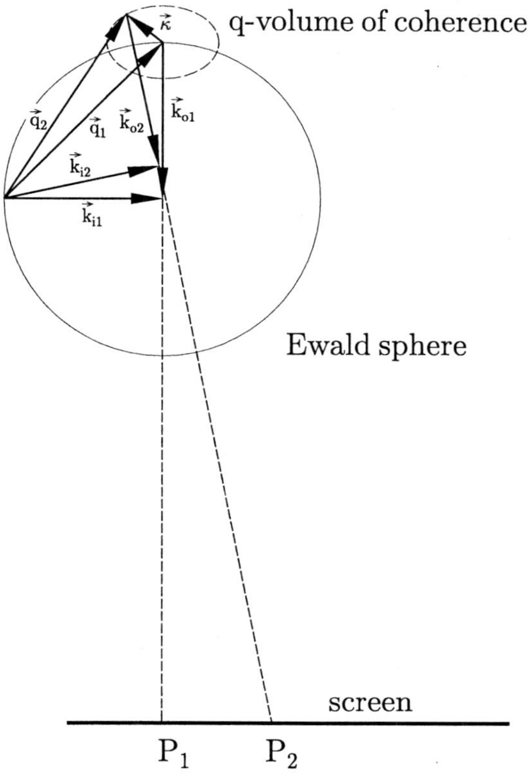

Fig. 4.7. The definition of the q-volume of coherence and its experimental meaning. Further explanation see text.

the q-volume of coherence cuts off the Ewald sphere. This has an important consequence: Even if the wave vector $\boldsymbol{q}$ of a refractive index wave does not end up exactly on the Ewald sphere, one will observe scattered light if at least the q-volume of coherence cuts the Ewald sphere.

4.7 The Temporal Change of the Speckle Pattern

4.7.1 The Motion of the Pattern

We discussed in Sect. 4.5 that a plane refractive index wave can give rise to a scattered light wave in a particular direction, provided the wave vector of this wave and the k-vector of the illuminating light beam obey the condition (4.37). In that case, one would observe a bright spot on the observation screen in Fig. 4.2. If now the amplitude $K_o(t)$ (c.f. equation (4.35)) changed in time, the amplitude of the scattered light would change in the same manner. A change in the phase ϕ of the refractive index wave, which corresponds to a spatial shift of this wave, would not produce any change in the observed intensity of the scattered light.

Let us now assume that for some reason the $\boldsymbol{q}$-vector of this refractive index wave changes. A simple rotation of the vessel containing the medium leads to such a change, provided the rotation is not about an axis parallel to $\boldsymbol{q}$. If this change is such that the condition (4.37) remains fulfilled, which means that the vector $\boldsymbol{q}$ stays on the Ewald sphere, one would continue to observe the spot of scattered light, but its location on the screen would likewise change. The spot would disappear, however, in case the condition (4.37) does not remain valid.

We will point to the fact, that the spot of scattered light in the above experiment will only move or disappear, if the refractive index wave remains a plane wave, regardless of whether its amplitude K_o, phase ϕ or wave vector $\boldsymbol{q}$ change in time. If, however, the wave becomes distorted so that it is no longer a plane wave, but the surfaces of constant phase - the so called phase surfaces - become curved, then the spot of scattered light will generally extend and assume an irregular shape. This can be explained in the following way: A distorted wave can be considered as a superposition of many *plane* waves with different wave vectors, amplitudes and phases. A certain fraction - those for which (4.37) is fulfilled - of these plane waves will have wave vectors which give rise to scattered light.

We are now going to show that a plane refractive index wave, which is materially fixed to a flowing fluid, always remains a plane wave when the velocity field of the flow has the form (4.2).

In order to calculate the temporal change of the refractive index wave in such a flowing medium, let us first look at this wave at a fixed time t_o:

$$K(\boldsymbol{r}, t_o) = K_o \cos(\boldsymbol{q}_o \cdot \boldsymbol{r} + \phi_o). \tag{4.58}$$

In order to know what the wave looks like at a later time, we have to calculate at what location $\boldsymbol{r}(t)$ an 'atom' of the fluid is found at a later time t, if it was at the arbitrary location $\boldsymbol{r}_o = \boldsymbol{r}(t_o)$ at time t_o. This was already calculated in Sect. 4.3:

$$\boldsymbol{r}(t) = \hat{R}(t - t_o) \cdot \boldsymbol{r}_o + \hat{S}(t - t_o) \cdot \boldsymbol{v}_o. \tag{4.59}$$

By multiplying this equation from the left hand side with the inverse of the tensor $\hat{R}(t - t_o)$ (c.f. equation 3.95): the resulting equation can be resolved with respect to $\boldsymbol{r}_o$ and yields

$$\boldsymbol{r}_o = \hat{R}^{-1}(t - t_o) \cdot \boldsymbol{r}(t) - \hat{R}^{-1}(t - t_o) \cdot \hat{S}(t - t_o) \cdot \boldsymbol{v}_o. \tag{4.60}$$

Since the optical inhomogeneity is assumed to be materially fixed to the fluid, we can conclude that the inhomogeneity, which was at time t_o at the location $\boldsymbol{r}_o$, is at time t at the location $\boldsymbol{r}(t)$. This can be expressed by the following equation:

$$K(\boldsymbol{r}(t),t) = K(\boldsymbol{r}_o,t_o). \tag{4.61}$$

We are now going to replace $\boldsymbol{r}_o$ by $\boldsymbol{r}(t)$ by use of equation (4.60) and obtain:

$$\begin{aligned} K(\boldsymbol{r}(t),t) &= K_o\cos(\boldsymbol{q}_o\cdot\boldsymbol{r}_o+\phi_o) \\ &= K_o\cos\big(\boldsymbol{q}_o\cdot\hat{R}^{-1}(t-t_o)\cdot\boldsymbol{r}(t) - \\ &\quad \boldsymbol{q}_o\cdot\hat{R}^{-1}(t-t_o)\hat{S}(t-t_o)\cdot\boldsymbol{v}_o+\phi_o\big). \end{aligned} \tag{4.62}$$

The last equation does no longer depend on $\boldsymbol{r}_o$ and is valid for all $\boldsymbol{r}(t)$. Therefore, we replace $\boldsymbol{r}(t)$ by $\boldsymbol{r}$ and finally obtain

$$\begin{aligned} K(\boldsymbol{r},t) = K_o\cos\big(&\boldsymbol{q}_o\cdot\hat{R}^{-1}(t-t_o)\cdot\boldsymbol{r} - \\ &\boldsymbol{q}_o\cdot\hat{R}^{-1}(t-t_o)\hat{S}(t-t_o)\cdot\boldsymbol{v}_o+\phi_o\big). \end{aligned} \tag{4.63}$$

This equation can be written in a much simpler form. For that purpose we rewrite the first term of the argument of the cosine on the right hand side with the help of equation (3.85):

$$\boldsymbol{q}_o\cdot\hat{R}^{-1}(t-t_o)\cdot\boldsymbol{r} = (\hat{R}^{-T}(t-t_o)\cdot\boldsymbol{q}_o)\cdot\boldsymbol{r}. \tag{4.64}$$

In this equation we used the relation $\left(\hat{R}^{-1}\right)^T = \hat{R}^{-T}$. Now we introduce the following quantities:

$$\begin{aligned} \boldsymbol{q}(t) &= \hat{R}^{-T}(t-t_o)\cdot\boldsymbol{q}(t_o) \\ \psi(t) &= \boldsymbol{q}(t_o)\cdot\hat{R}^{-1}(t-t_o)\hat{S}(t-t_o)\cdot\boldsymbol{v}_o+\phi_o. \end{aligned} \tag{4.65}$$

Here $\boldsymbol{q}_o$ was replaced by $\boldsymbol{q}(t_o)$, since $\boldsymbol{q}_o$ was defined as the wave vector of the refractive index wave at time t_o. Equation (4.63) can now be written as

$$K(\boldsymbol{r},t) = K_o\cos\big(\boldsymbol{q}(t)\boldsymbol{r}+\psi(t)+\phi_o\big). \tag{4.66}$$

So we see that the plane refractive index wave remains a plane wave, but with changed wave vector and also changed phase. The second equation of (4.65) shows that the phase is dependent on both, the velocity $\boldsymbol{v}_o$ and $\hat{\Gamma}$, the new wave vector only depends on $\hat{\Gamma}$.

We will summarize those results, which are important for us:

- A spatially constant velocity of the scattering medium leads to a time dependent phase shift of the refractive index wave. As a consequence, the light intensity of the scattered light is not affected, only its phase.
- A velocity gradient alters a plane refractive index wave in such a way, that only its wave vector changes in time. In general, this causes a spatial change of the scattered light intensity. In addition, the phase is usually also affected by a velocity gradient.

A beautiful demonstration of the preservation of a plane wave, but with altered wave vector, within a homogeneous flow is given in [14].

Equation (4.65) can now be used to describe quantitatively the 'motion' of the scattered light. If, for instance, $\boldsymbol{q}(t_o)$ denotes a bright spot on the observation screen, then the vector $\boldsymbol{q}(t)$ from equation (4.65) gives the location of this spot at the later time t. This spot can only be observed, however, if $\boldsymbol{q}(t)$ remains on the Ewald sphere. The situation is analogous to the one depicted in Fig. 4.5 if $\boldsymbol{q}(t_o)$ is identified with $\boldsymbol{q}_1$ and $\boldsymbol{q}(t)$ with $\boldsymbol{q}_2$. If, on the other hand, the wave vector of the refractive index wave changes so that it does not remain on the Ewald sphere, then the bright spot will disappear. Of course, this disappearing is not abrupt. As we discussed at the end of the last section the finite size of the scattering volume leads to a finite size of the q-volume of coherence. A motion of the q-volume of coherence away from the Ewald sphere leaves the intensity on this sphere correlated with the initial intensity, as long as the motion is less than about half the diameter of this volume. So one will observe a gradual change of the considered bright spot. Generally, it changes shape and intensity.

The motion of the intensity pattern on a screen resulting from scattering of a random distribution of particles within a flow obeys the same laws as those just derived. The only difference is that there will be many scattering specks on the screen instead of just one. The random optical inhomogeneity can always be considered as a superposition of plane refractive index waves with randomly distributed amplitudes, wave vectors, and phases. All those waves with wave vectors on the Ewald sphere contribute to the scattered light. If, due to a velocity gradient within the fluid, the temporal change of the $\boldsymbol{q}$- vectors in a particular region leads away from the Ewald sphere, the corresponding specks disappear. But, of course, the screen does not get dark because the $\boldsymbol{q}$-vectors of other refractive index waves move towards the Ewald sphere and new specks on the screen appear. One therefore observes a temporal change of the speckle pattern, which is known as 'boiling' (see Sect. 3.5).

Let us now calculate the velocity of the speckle pattern :

$$\begin{aligned}\dot{\boldsymbol{q}}(t) &= \frac{d}{dt}\hat{R}^{-1}(t-t_o)\cdot\boldsymbol{q}(t_o) \\ &= \frac{d}{dt}\exp\left(-\hat{\Gamma}^T(t-t_o)\right)\cdot\boldsymbol{q}(t_o) \\ &= -\hat{\Gamma}^T\cdot\exp\left(-\hat{\Gamma}^T(t-t_o)\right)\cdot\boldsymbol{q}(t_o)\end{aligned} \tag{4.67}$$

Setting $t_o = t$ in this equation leads to

$$\dot{\boldsymbol{q}} = -\hat{\Gamma}^T\cdot\boldsymbol{q}. \tag{4.68}$$

In this last equation we did no longer write the time dependence of $\boldsymbol{q}$ explicitly.

The meaning of the vector $\boldsymbol{k}_o$ is more intuitively obvious than that of the vector $\boldsymbol{q}$. We observe that the vector $\boldsymbol{k}_i$ is constant. That means

$$\dot{\boldsymbol{q}} = \frac{d\boldsymbol{q}}{dt} = \frac{d(\boldsymbol{k}_i - \boldsymbol{k}_o)}{dt} = -\frac{d\boldsymbol{k}_o}{dt} = -\dot{\boldsymbol{k}}_o \tag{4.69}$$

and we can therefore write

$$\dot{\boldsymbol{k}}_o = \hat{\Gamma}^T \cdot (\boldsymbol{k}_i - \boldsymbol{k}_o) \, . \tag{4.70}$$

In this equation, only the component of $\dot{\boldsymbol{k}}_o$ that is perpendicular to $\boldsymbol{k}_o$ describes a motion of the speckle pattern. We will denote this component by the subscript $\perp$:

$$\dot{\boldsymbol{k}}_{o\perp} = \left[\hat{\Gamma}^T \cdot (\boldsymbol{k}_i - \boldsymbol{k}_o)\right]_\perp \, . \tag{4.71}$$

The other component, which is parallel to $\boldsymbol{k}_o$, produces boiling of the pattern. The general case, where $\dot{\boldsymbol{k}}_o$ is not purely perpendicular to $\boldsymbol{k}_o$, will be called by us 'oblique' speckle motion.

It is advantageous for the practical application of the last relationship, which represents the basis of all measurement methods described in this book, to divide the entire equation by $|\boldsymbol{k}|$

$$\frac{\dot{\boldsymbol{k}}_{o\perp}}{|\boldsymbol{k}_o|} = \left[\hat{\Gamma}^T \cdot (\boldsymbol{e}_i - \boldsymbol{e}_o)\right]_\perp \, . \tag{4.72}$$

$\boldsymbol{e}_i$ and $\boldsymbol{e}_o$ are unit-vectors, which simply indicate the directions of the illuminating laser beam and the direction in which the scattered light is observed. Obviously, the right hand side of this equation no longer depends on the wavelength of the light, and consequently the same must be true for the left hand side.

Due to the property of $\dot{\boldsymbol{k}}_{o\perp}$ to be perpendicular to $\boldsymbol{k}_o$, we conclude that the vector

$$\boldsymbol{\omega}_p = \frac{\dot{\boldsymbol{k}}_{o\perp}}{|\boldsymbol{k}_o|} \tag{4.73}$$

just represents the angular velocity $\boldsymbol{\omega}_p$ of the scattered light intensity field around a point within the scattering volume.

The significance of equation (4.71) for the determination of components of the velocity gradient tensor $\hat{\Gamma}$ will now be explained. Let us denote the direction of the scattering vector $\boldsymbol{k}_i - \boldsymbol{k}_o$ as direction l. The velocity of the pattern shall be measured in direction j (for the notation see equation (3.65)).

We then have

$$\dot{k}_{oj} = \Gamma^T_{jl} \cdot (k_{il} - k_{ol}) \tag{4.74}$$

or

$$\Gamma^T_{jl} = \Gamma_{lj} = \frac{\partial u_l}{\partial x_j} = \frac{\dot{k}_{oj}}{(k_{il} - k_{ol})} \, . \tag{4.75}$$

That means:

- The scattering vector $\boldsymbol{q} = \boldsymbol{k}_i - \boldsymbol{k}_o$ denotes the velocity component, of which the derivative is measured
- The direction in which the pattern velocity is measured is the same as the differentiation direction of the flow velocity, which determines the velocity gradient component responsible for the pattern motion in that direction.

It is not difficult to calculate for any scattering geometry and a specified co-ordinate system the linear combination of velocity gradient components which is measured. Let us for that purpose introduce the unit-vector $\boldsymbol{e}_P$, which denotes the direction in which the motion of the pattern is measured. (n.b. that $\boldsymbol{e}_P$ is perpendicular to $\boldsymbol{k}_o$!). Let us furthermore denote the unit-vector parallel to $\boldsymbol{q}$ by $\boldsymbol{e}_q$. The Cartesian components of these two unit-vectors are called direction cosines and we write:

$$\boldsymbol{e}_q = \begin{pmatrix} \alpha_q \\ \beta_q \\ \gamma_q \end{pmatrix} \tag{4.76}$$

$$\boldsymbol{e}_P = \begin{pmatrix} \alpha_P \\ \beta_P \\ \gamma_P \end{pmatrix} .$$

The size of the component of $\dot{\boldsymbol{k}}_{o\perp}$ which is responsible for the pattern motion in direction $\boldsymbol{e}_P$ is then simply given by the scalar product of these two vectors. I.e. by multiplication of equation (4.72) with $\boldsymbol{e}_P$ we obtain the desired linear combination:

$$\frac{\dot{\boldsymbol{k}}_o \cdot \boldsymbol{e}_P}{|\boldsymbol{k}_o|} = \boldsymbol{e}_P \cdot \hat{\Gamma}^T \boldsymbol{e}_q = e_{Pi} \frac{\partial u_j}{\partial x_i} e_{qj} \tag{4.77}$$

where we used the summation convention. If we now introduce the quantitiy

$$u_q = \boldsymbol{u} \cdot \boldsymbol{e}_q = \alpha_q u_x + \beta_q u_y + \gamma_q u_z \tag{4.78}$$

we can rewrite the last equation in the form

$$\frac{\dot{\boldsymbol{k}}_o \cdot \boldsymbol{e}_P}{|\boldsymbol{k}_o|} = \alpha_P \frac{du_q}{dx} + \beta_P \frac{du_q}{dy} + \gamma_P \frac{du_q}{dz} , \tag{4.79}$$

the right hand side of this equation is just the desired linear combination of velocity gradient components.

So far we neglected the Brownian motion of the scattering particles, which changes the particle paths in the flowing fluid to random paths. Fortunately, it is not difficult to calculate the average paths in the fluid:

$$\langle \boldsymbol{r}(t) \rangle = \int P_c(\boldsymbol{r}; t | \boldsymbol{r}_o; t_o) \boldsymbol{r} dV \tag{4.80}$$

Insertion of $\int P_c(\boldsymbol{r};t|\boldsymbol{r}_o;t_o)$ from equation (4.10) yields

$$\langle \boldsymbol{r}(t)\rangle = (4\pi D)^{-\frac{3}{2}}\left[\det(\hat{A})\right]^{-\frac{1}{2}} \cdot \tag{4.81}$$
$$\int \exp\left(\frac{1}{4D}(\boldsymbol{r}-\hat{R}\cdot\boldsymbol{r}_o-\hat{S}\cdot\boldsymbol{v}_o)\cdot\hat{A}^{-1}\cdot(\boldsymbol{r}-\hat{R}\cdot\boldsymbol{r}_o-\hat{S}\cdot\boldsymbol{v}_o)\right)\cdot\boldsymbol{r}dV.$$

The integration can be carried out analytically and we obtain:

$$\langle \boldsymbol{r}(t)\rangle = \hat{R}(t-t_o)\cdot\boldsymbol{r}_o + \hat{S}(t-t_o)\cdot\boldsymbol{v}_o. \tag{4.82}$$

So we end up with the result that the average particle path in a homogeneous flow is not altered by diffusion. This obviously means that the average motion of the speckle pattern is not altered by diffusion either.

4.7.2 The 'Boiling' of the Speckle Pattern

As we showed in the last section, the Brownian motion does not lead to an alteration in the speckle motion, but it causes a random temporal change of the pattern, which is superimposed on the motion and renders the measurement of speckle motion quite difficult. Among the several mechanisms, which lead to speckle boiling, the Brownian motion is usually the one of least significance. Since the Brownian motion is always present in experiments of the type we are going to consider, and since it leads to a peculiar property of the temporal correlation function of the scattered light, we are going to calculate this correlation function.

We will do that for the special case of a fluid at rest, i.e. one with $\hat{\Gamma}=0$ and $\boldsymbol{v}_o=0$. Let us then write down the correlation function of the electric field for this case:

$$\langle E(\boldsymbol{q},t_o)E^*(\boldsymbol{q}+\boldsymbol{\kappa},t_o+t)\rangle = E_o^2\int\int G(\boldsymbol{r})G(\boldsymbol{r}')P^2(\boldsymbol{r},\boldsymbol{r}';t,t_o) \tag{4.83}$$
$$\exp(i\boldsymbol{q}\cdot\boldsymbol{r}')\exp(i(\boldsymbol{q}+\boldsymbol{\kappa})\cdot\boldsymbol{r})\exp\left(-i\omega(t-t_o)\right)dVdV'.$$

The joint probability density can according to equation (3.27) be expressed by the conditional probability density (4.10):

$$P_c(\boldsymbol{r},\boldsymbol{r}_o;t,0) = \frac{1}{(4\pi D)^{3/2}}\cdot\exp\left(-\frac{(\boldsymbol{r}-\boldsymbol{r}_o)^2}{4D|t|}\right)\frac{1}{V}. \tag{4.84}$$

Here V is the size of the considered volume and we put $t_o=0$. With the approximation $G(\boldsymbol{r})=\text{const}=1$ we then obtain

$$\langle E(\boldsymbol{q},t_o)E^*(\boldsymbol{q}+\boldsymbol{\kappa},t_o+t)\rangle = \tag{4.85}$$
$$E_o^2\exp\left(-\frac{D}{4}\cdot(2\boldsymbol{q}+\boldsymbol{\kappa})^2\cdot|t|\right).$$

(It should be noted that the first integration extends over a finite volume of size V, the second one must be done over an infinite volume). As can be seen, in this case the time dependence of the correlation function is exclusively determined by the diffusion constant. This is not surprising, because without diffusion the particles would not move at all and the scattered light would behave likewise. In the more general case of $\hat{\Gamma} \neq 0$ and $\boldsymbol{v}_o \neq 0$ the exponential factor, which is determined by the diffusion, depends also on $\hat{\Gamma}$. This leads to a faster decrease of the correlation function compared to a fluid at rest.

It seems strange now that the correlation function has - as a function of t - a cusp at $t = 0$, i.e. it is not differentiable there. This must be taken into account for measurement schemes which are based on temporal derivatives of the scattered light intensity (see Chap. 9).

A further mechanism that leads to speckle boiling - generally much faster than the diffusion - is the random flux of particles into and out of the scattering volume. Obviously, scattered light cannot stay correlated for longer than the longest period an arbitrary particle stays within the scattering volume. After that period, all particles within this volume are replaced by other ones, the locations of which are completely independent of those the other particles had before.

In general, the in-/outflux of particles is determined predominantly by the velocity $\boldsymbol{u}_o$. According to what we just said, the correlation time caused by this effect is the time the particles need to travel through the scattering volume, and this time is given by the diameter d_S of this volume in direction of $\boldsymbol{u}_o$ and the absolute value of $\boldsymbol{u}_o$:

$$\tau_V = \frac{d_S}{|\boldsymbol{u}_o|}. \tag{4.86}$$

It is therefore possible to reduce this contribution to the speckle boiling by choosing a scattering volume with a greater diameter in the direction of $\boldsymbol{u}_o$. The influx and outflux of particles is also caused by the velocity gradient itself. However, the contribution of it to speckle boiling is usually very small. Only if the average velocity is very small and the gradient large, which usually occurs near walls, one might observe a considerable contribution due to the gradient.

But the velocity gradient is the main source of boiling for an other reason: In all those cases where, due to the velocity gradient, the temporal derivative $\dot{\boldsymbol{k}}_o$ is not perpendicular to $\boldsymbol{k}_o$, a boiling pattern is produced. In turbulent flows with randomly changing velocity gradients that will always be the case, but even in laminar flows it will only in rare occasions be possible to choose the scattering geometry so that the two vectors $\dot{\boldsymbol{k}}_o$ and $\boldsymbol{k}_o$ are always perpendicular to each other.

A possibility to quantify this boiling produced by oblique speckle motion exists only if both, the direction of $\dot{\boldsymbol{k}}_o$ and the spatial aperture function (see discussion following equation 4.29), are known.

4.7.3 'Delusive' Speckle Motion

At the end of this chapter we have to discuss an effect which can influence both, speckle motion direction and velocity, in such a way that equations (4.70) and (4.71) are no longer valid.

This effect will not be treated quantitatively here since it is advisable to prevent its occurrence by a properly devised experimental set-up. Figure refB9 a) shows schematically the set-up of a scattering experiment which is more elaborate than the one shown in Fig. 4.2. However, this set-up allows for rather accurate control of the size and shape of the scattering volume. This will be shown in Sect. 5.2 by means of the similar set-up depicted in Fig. 5.3. Two aspects are important here: Firstly, light exclusively from the region denoted 'scattering volume' reaches the observation screen. The other aspect has already been discussed in connection with Fig. 4.2. Nevertheless, we will state it once more because of its significance: The intensity at a point on the screen is determined exclusively by *one* component of the scattered light with a definite wave vector. The cigar shaped scattering volume is produced in the set-up by the thin laser beam and the large diaphragm B. In part b) of this figure, the resulting q-volume of coherence is indicated, which also looks cigar shaped; spatially it has the form of a disk.

Let us assume that at time $t = 0$ a refractive index wave with the wave vector $\boldsymbol{q}_o$ exists in the scattering volume. Then one observes on the observation screen just one bright line - perpendicular to the plane of the drawing - at a location denoted by P_o. We will now assume that the velocity field of the flow is such that the wave vector of the refractive index wave moves in the direction parallel to $\boldsymbol{k}_o$. Then this wave vector has at time $t = t_1$ the value $\boldsymbol{q}_1$. According to what we said above, no pattern motion whatsoever should occur in this situation, because the vector $\boldsymbol{k}_o$ changes its length only and its temporal derivative is therefore parallel to $\boldsymbol{k}_o$. This should result simply in a boiling of the pattern.

As a matter of fact, one observes the bright line although with lower intensity also at the later time $t = t_1$ at the location P_1 on the observation screen. The reason for that is that the q-volume of coherence cuts the Ewald sphere at the corresponding location.

It is thus obvious that in this case the pattern of the scattered light performs a motion that should not be present, if the theory outlined above were sufficient to describe pattern motion.

However, it is possible to get rid of this 'delusive' speckle motion with the choice of a scattering volume which has rotational symmetry about an axis parallel to $\boldsymbol{k}_o$. It is not difficult to make that clear with the help of Fig. 4.8. If in part b) of this figure the q-volume of coherence were disc shaped, but with the short axis in the direction of $\boldsymbol{k}_o$, no motion of the bright spot could be observed for the indicated change of the refractive index wave. The only temporal change of the pattern would be a disappearing of the bright line. It is also obvious that for this q-volume of coherence, the motion of the

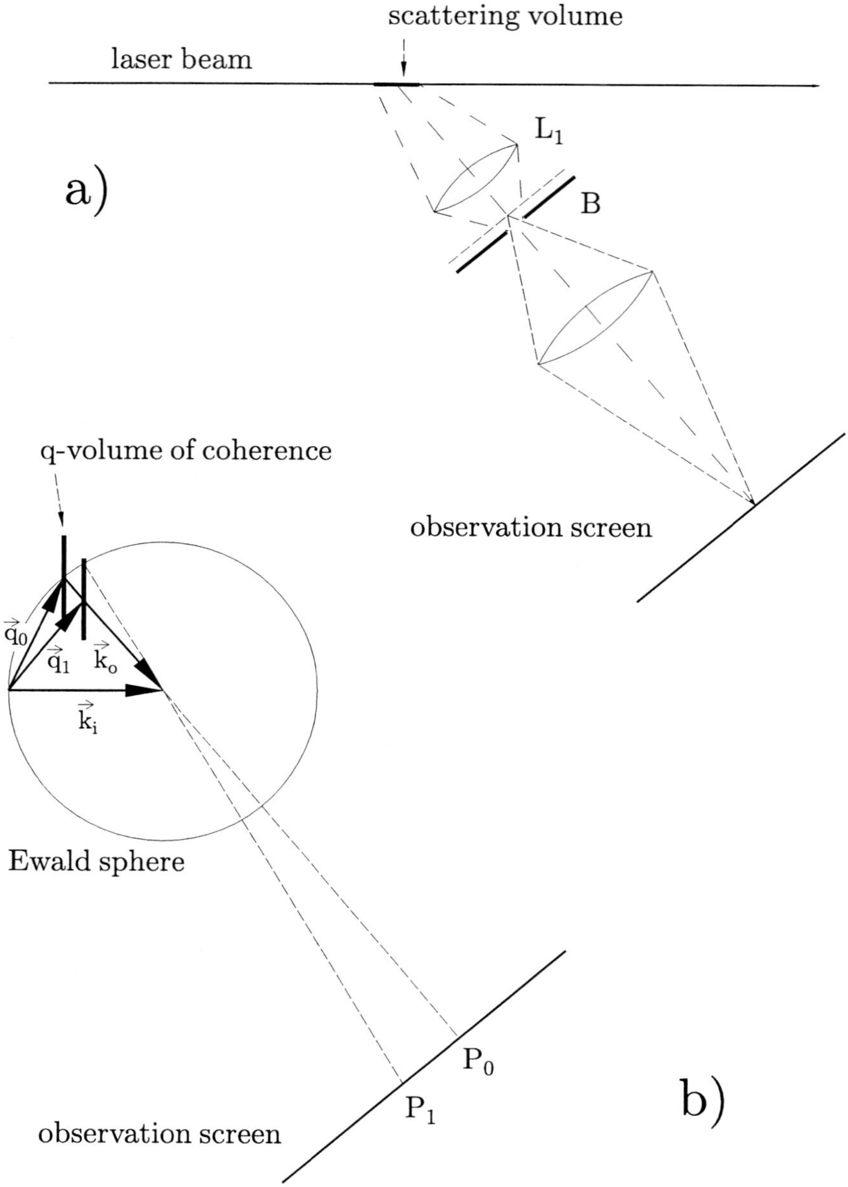

Fig. 4.8. The origin of delusive speckle motion. Part a) The choice of a large diaphragm B leads to a cigar shaped scattering volume. The resulting q-volume of coherence is shown in part b). Part b) shows the mechanism leading to delusive speckle motion: If a refractive index wave changes in a way that its wave vector develops from the value $\boldsymbol{q}_o$ to $\boldsymbol{q}_1$, one observes a motion of the bright spot on the screen from point P_0 to P_1, although the q-volume of coherence travels perpendicular to the Ewald sphere, (in the depicted case towards its centre).

coherence area on the Ewald sphere is just the projection of the motion of the vector $\dot{\boldsymbol{k}}_o$ onto the Ewald sphere, which is $\dot{\boldsymbol{k}}_{o\perp}$ of equation (4.72). Obviously, required rotational symmetry can be achieved especially easily for right angle scattering, i.e. when the vectors $\boldsymbol{k}_o$ and $\boldsymbol{k}_i$ are perpendicular to each other.

In cases where only one particular component of the pattern velocity is measured, the requirement concerning the symmetry of the scattering volume in order to prevent delusive pattern motion may be less strict. If, for instance, in the experiment shown in Fig. 4.8 one is interested in the pattern velocity component perpendicular to the plane of the drawing only, no delusive pattern motion occurs and, therefore, there is no need for right angle scattering.

5. Physical Basis: Experimental Part

5.1 The Illuminating Laser Beam

A laser beam can be characterized by a number of parameters, which are of different interest for the measurement techniques we are going to consider here. Besides the wavelength, it is especially the parameters describing the profile of the beam, which are of utmost importance. Most lasers nowadays produce beams in the so called TEM_{oo}-mode. Such beams are rotationally symmetric, or may also have an elliptic profile, and the intensity across the beam can be described by a two-dimensional Gaussian function.

Laser beams always diverge when they leave the laser resonator. This means that they do not fulfil the necessary requirement for the measurement techniques described here, which is to represent a plane light wave (see Sect. 4.5). It is possible to achieve a close approximation of a plane light wave within a limited region of space by focusing the beam. Waves of the type shown in Fig. 5.1 are produced by such focusing using a convex lens. The phase surface at the location where the wave has its smallest lateral extension - the so called beam waist - is exactly plane, whilst the curvature of the phase surfaces increases with the distance from that beam waist. It should be stressed that the location of the beam waist and that of the focus according to ray optics do *not* generally coincide. This is especially marked if the divergence of the beam is fairly small. Consequently, as far as our measurement technique is concerned one will always choose a scattering volume with the beam waist at its centre.

Apart from semiconductor lasers, the beam profile of lasers is rotationally symmetric. The difference in the two extreme beam diameters for those elliptic profile lasers is usually fairly large. In some cases it may be of advantage to use this beam profile, but often one will prefer a circular beam. By use of cylindrical lenses it is possible to form circular beams from elliptic beams, indeed some semiconductor lasers have the optics for such beams already integrated.

The measurement techniques discussed in this text have no specific requirements with regards to the coherence length. Generally, this does not need to exceed significantly the diameter of the scattering volume in direction of the beam axis.

If the laser beam is not of the TEM_{oo}-type, one can get rid of the disturbing beam components by filtering the beam with a mode filter. However, this procedure normally results in a significant reduction of the beam power.

TEM_{oo}-beams are usually described sufficiently accurate by the model of 'Gaussian beam' [12]. Let us consider a rotationally symmetric TEM_{oo} beam which travels in the z-direction. Such a beam can be described in an optically homogeneous medium (a medium with a constant refractive index) by the following parameters:

- λ: the wavelength
- z_o: the location of the smallest radius of the beam, the so called beam waist
- w_o: the size of the beam radius at the beam waist.

These parameters determine the other interesting quantities of the beam as:

- $w(z)$: the radius of the beam at an arbitrary location z on the beam axis
- ϕ_o: the angle of divergence
- $R(z)$: the radius of curvature of the phase surfaces.

The nature of the quantities described may be easily understood by examining Fig. 5.1.

The connections between these quantities are listed below:

$$\begin{aligned} \tan(\phi_o) &= \frac{\lambda}{\pi w_o} \\ w(z) &= w_o \sqrt{1 + \frac{\lambda^2 (z - z_o)^2}{\pi^2 w_o^4}} \\ R(z) &= (z - z_o) \cdot \left(1 + \frac{\pi^2 w_o^4}{\lambda^2 (z - z_o)^2}\right) . \end{aligned} \tag{5.1}$$

The beam radius w at the location z on the beam axis is defined such that the intensity at a distance w away from the beam axis is smaller by a factor of $1/e^2$ than on the axis.

It can be seen from the last equation that, at the beam waist, the radius of curvature of the phase surface is infinite, i.e. that the phase surface is plane. The radius of curvature then decreases with the distance from the location of the beam waist and reaches a minimum value at $z = z_o \pm \pi w_o^2/\lambda$. For increasing distances, the radius of curvature again increases, i.e. the phase surfaces tend to become plane again. In Fig. 5.1 the loci of strongest curvature are marked by dots. Of all properties of Gaussian beams, this one is the most important for the measurement techniques described here, because they require a *plane* wave as the illuminating light wave.

After a transition of a Gaussian beam from one medium with refractive index n_1 to another one with a different refractive index $n_2 \neq n_1$, the beam parameters are changed. Generally, a Gaussian beam stays Gaussian only if its angle of divergence is small and the beam axis is perpendicular to the plane separating the two media. If we denote the location where this plane cuts the beam axis by z_g, and provide all parameters with an index denoting the medium it refers to, we get the following relations:

$$\begin{aligned}(\phi)_2 &= \frac{n_1}{n_2}\cdot(\phi)_1 \\ (w_o)_1 &= (w_o)_2 \\ (z_o)_2 &= \frac{n_2}{n_1}\cdot((z_o)_1 - z_g) + z_g .\end{aligned} \tag{5.2}$$

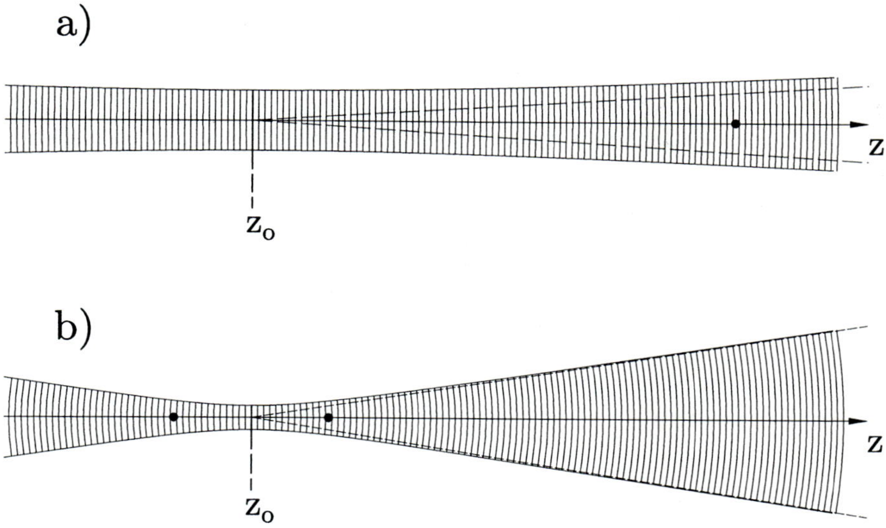

Fig. 5.1. Two Gaussian beams near the beam waist. a) $w_o = 5\lambda$, b) $w_o = 2\lambda$. Neighbouring phase surfaces are one wavelength apart from each other. The dots mark the locations of the phase surfaces with largest curvature. On the right hand side of the beam waists, lines are drawn which represent the limiting beams in the corresponding picture of geometric optics

It is worthwhile noting that the location z_o of a beam waist is often 'virtual', i.e. the beam does not actually reach this point. Therefore, the parameter z_o must be looked upon as a quantity describing beam properties within an optically homogeneous region, (i.e. one with constant refractive index), rather than an actual location.

If a Gaussian beam passes a plane parallel plate at right angles only the value of z_o is changed by

$$\Delta z_o = d\cdot\left(1 - \frac{n_1}{n_2}\right). \tag{5.3}$$

n_2 is the refractive index of the plate, n_1 is that of the surrounding medium.

If a beam does not cross a separation plane of two media at right angles, i.e. not at normal incidence, an awkward degradation of the beam results; the so called astigmatic distortion. The refraction at the separation plane leads to different divergence angles in two orthogonal planes, one of which is formed by the beam direction and the normal of the separation plane, and the other one by the plane which is perpendicular to it and also contains the beam axis. The result of these different divergence angles are different locations of the beam waists in these two planes; therefore, no location exists where the phase surfaces are plane. The astigmatic distortion increases with increasing angle of incidence, (i.e. increasing deviation from normal incidence), divergence angle and difference of the refractive indices. By the passage through a plane parallel plate, the astigmatism will furthermore increase with thickness of the plate. A theoretical analysis of the influence of this type of beam distortion on the pattern motion has not yet been performed.

By the application of lenses it is possible to shift the location of the beam waist, alter divergence angle and diameter of the beam, and to produce several beam waists if convex lenses are used.

Simple lenses always produce beams with spherical aberration (see Fig. 5.2), which, in spite of their rotational symmetry, have no plane phase surfaces in the region of the beam waist. One can circumvent this difficulty by use of achromatic lenses. They usually consist of two cemented lenses and are designed in a way that not only the wavelength dependence of the focal length is corrected - from where their name results - but also the spherical aberration. However, both aberrations are corrected to a certain degree only.

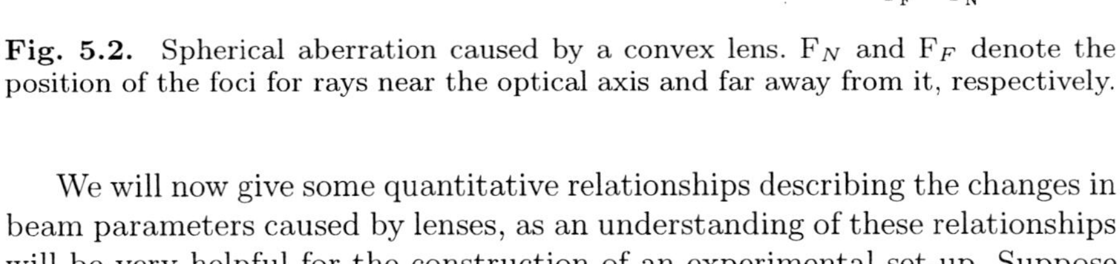

Fig. 5.2. Spherical aberration caused by a convex lens. F_N and F_F denote the position of the foci for rays near the optical axis and far away from it, respectively.

We will now give some quantitative relationships describing the changes in beam parameters caused by lenses, as an understanding of these relationships will be very helpful for the construction of an experimental set-up. Suppose one had a ray with a beam waist at z_o with radius w_o. If a lens with focal length f is placed in that beam at z_L, the beam waist will be shifted to z_1 with the modified radius w_1. With the abbreviation

$$d = z_L - z_o, \tag{5.4}$$

one obtains the following equations for the calculation of the modified beam parameters:

$$z_1 = z_L + f \frac{\frac{\pi^2 w_o^4}{\lambda^2 f^2} - \frac{d}{f}\left(1 - \frac{d}{f}\right)}{\left(1 - \frac{d}{f}\right)^2 + \frac{\pi^2 w_o^4}{\lambda^2 f^2}} \tag{5.5}$$

and

$$w_1 = \sqrt{\frac{w_o^2}{\left(1 - \frac{d}{f}\right)^2 + \frac{\pi^2 w_o^4}{\lambda^2 f^2}}} . \tag{5.6}$$

These equations were calculated within the framework of matrix optics outlined by Kogelnik and Li [12]. It should be noted that they are approximations, which give in most cases sufficiently accurate results, provided the angle of divergence of the beam is small. It is furthermore worth noting that these equations were derived for thin lenses. If thick lenses or systems of lenses (e.g. photographic lenses) are used the relationships become more complicated, since the influence of further parameters describing lens properties must be taken into account, e.g. the position of principal planes. For an estimate of beam parameters in an optical set-up the given equations will usually give sufficiently accurate results.

One must keep in mind that the locations of the beam waists, which one obtains by application of equation (5.5), do *not* coincide with the location of the focus one obtains in the framework of geometrical optics (see equation (5.23)). Let us for instance assume that there is an object at z_o (the notation refers to that introduced above). This object is to be imaged with a lens of focal length f. Then the picture of this object is *not* found at z_1. This discrepancy is due to the fact that the theory just mentioned comprises both refraction and diffraction, while in geometrical optics only refraction is considered.

5.2 The Scattering Volume

We introduced in Sect. 4.5 the spatial aperture function $G(\boldsymbol{r})$, which, roughly speaking, denotes the spatial region from which scattered light reaches the observation screen. In a set-up corresponding to Fig. 4.2, the scattering volume has the form of a long cylinder; its length is determined by the diameter of the lens and the cross-sectional area by the illuminating laser beam. Since the phase surfaces of the laser beam are only plane near the beam waist, one can conclude that this optical set-up is not suitable for our purpose.

It therefore makes sense to choose an optical set-up as the one shown schematically in Fig. 5.3. Here the lens L_1 images the laser beam onto the plane of the diaphragm B. The diameter of the scattering volume in direction of the laser beam can be controlled by the diameter of this diaphragm. If this

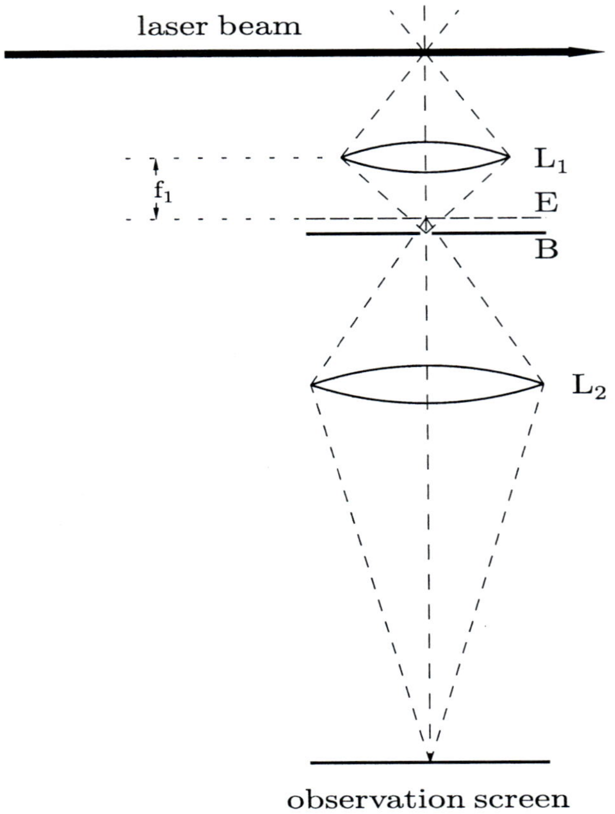

Fig. 5.3. Schematic of a scattering experiment which allows for control of the scattering volume. Its extension in direction of the laser beam is determined by the magnification caused by lens L_1 and the diameter of diaphragm B. Lens L_2 images the focal plane E of lens L_1 on the observation screen.

diameter is smaller than the image of the laser beam, then even the extension of the scattering volume in this direction is determined by this diaphragm.

The second lens L_2 is placed in such a way that it images the focal plane E of lens L_1 onto the screen S - through the opening of the diaphragm. The benefits of such an arrangement of optical elements are as follows: A plane wave component, which passes through the lens L_1, will be focused onto a point in the plane E. Each point of this plane is then imaged by the lens L_2 onto a point of the screen S. This set-up therefore allows for control of the scattering volume, and it focuses any plane wave component of the scattered light field to a particular point on the observation screen in the same way as the set-up shown in Fig. 4.2. For that reason, the relationships describes in Sect. 4.5 are also valid for this new set-up.

The best way to visualize the spatial aperture function $G(\boldsymbol{r})$ is to think of a particle that can be moved in a controlled way within the scattering volume. The relative value of the aperture function is then given by the square root of the intensity of the scattered light integrated over the observation screen. The measurement techniques we are going to consider will not depend on the exact knowledge of this function. Therefore, we are not going to actually calculate it. Mainly, it is the extensions in three orthogonal directions and certain symmetry properties that are of importance. For instance, the laser beam can be distorted into an elliptical shape by introducing a cylindrical lens. The resulting disk shaped scattering volume is necessary for one of the measurement techniques described in Chap. 8.

5.3 The Light Scattering Particles

Apart from the method explained in Chap. 6, the forced Rayleigh scattering, the measurement methods described here require the seeding of the fluid under investigation with light scattering particles. The optical inhomogeneity caused by density fluctuations of pure fluids is usually so small that the intensity of the scattered light is insufficient for our purposes.

It is very important that the added particles be spherical, otherwise the intensity of the scattered light would not only depend on the position of the particles - as is needed for the application of our theory - but also on their orientation in space. Unfortunately, this orientation depends on the local velocity gradient, which generally causes a rotation of the particles. Therefore, this property does not only contribute to the boiling of the speckle pattern, but also to its motion. This was demonstrated experimentally by Frish and Webb [7].

For our investigation of water flows we used, with good success, small latex particles of $\approx$ 200 nm diameter. Adding a small amount of detergent (Extran) to the water delays the agglomeration of the particles for up to a few weeks. Particle agglomerates are no longer spherical and must therefore be removed. The optimal particle concentration depends on their scattering power and the length of the light path within the fluid. We are now going to calculate it assuming that the particles scatter the light but do not absorb it. Any possible absorption is assumed to be caused by the fluid, which will be included in the calculations.

Along a small path dl within the fluid, the fraction dl/Λ_s is scattered by the particles and the fraction dl/Λ_a is absorbed by the fluid. The two processes are quantified by the two constants Λ_s and Λ_a, the so called scattering length and absorption length, respectively. If the light intensity at the beginning of a small path ds is I, it will have reduced by

$$dI = -I\frac{dl}{\Lambda_s} - I\frac{dl}{\Lambda_a} \tag{5.7}$$

at the end of this path. This equation can easily be integrated and one obtains:

$$I(l) = I_o \exp\left(-\frac{l}{\Lambda_s} - \frac{l}{\Lambda_a}\right). \tag{5.8}$$

I_o is in this equation the light intensity at an arbitrary location and $I(l)$ the one a distance l away in the direction of the traveling light. From this equation, the physical meaning of the two parameters becomes obvious: They denote the path length which is necessary for the light intensity to be reduced to 1/e by the corresponding mechanism (scattering or absorption). Large values of these parameters therefore denote small scattering or absorption power.

Let us denote the path of the light from the point where it enters the scattering fluid to the scattering volume by l_s. The amount of light scattered in the scattering volume is, according to equation (5.7), proportional to d_s/Λ_s, where d_s is the extension of the scattering volume in the direction of the illuminating light beam. However, this is only a sufficient approximation if the diameter of the scattering volume is small compared to Λ_s, which will usually be the case. If the path of the light from the scattering volume up to the location where it leaves the fluid is denoted by l_a, one gets the following expression for the scattered light intensity.

$$\begin{aligned} I_s &= I_o \exp\left(-\frac{l_s}{\Lambda_s} - \frac{l_s}{\Lambda_a}\right) \cdot \frac{d_s}{\Lambda_s} \cdot \exp\left(-\frac{l_a}{\Lambda_s} - \frac{l_s}{\Lambda_a}\right) \\ &= I_o \exp\left(-\frac{L}{\Lambda_s} - \frac{L}{\Lambda_a}\right) \cdot \frac{d_s}{\Lambda_s}, \end{aligned} \tag{5.9}$$

In the last equation L denotes the entire light path $l_s + l_a$. Differentiating this equation with respect to Λ_s yields

$$\frac{\partial I_s}{\partial \Lambda_s} = I_o \exp\left(-\frac{L}{\Lambda_s} - \frac{L}{\Lambda_a}\right) \cdot \left[\frac{L}{\Lambda_s^3} - \frac{1}{\Lambda_s^2}\right] \cdot d_s. \tag{5.10}$$

The maximum of $I - s$ occurs if this equation is zero which will happen for $\Lambda_s = L$. This means that in case that no light absorption in the fluid occurs the concentration of the scattering particles should be such that the light on its entire path through the scattering fluid is reduced to one e-th.

Measurement errors will be introduced in almost all light scattering experiments by multiple scattering. Some of the methods described below react very sensitively to the occurrence of multiply scattered light, which also contributes to the coherent false light (see Sect. 5.5). If there is an appreciable amount of multiply scattered light, a halo around the illuminating laser beam can be observed. As quantitative treatments of the influence of multiply scattered light on the obtained results have not yet been performed, it is advisable to rule out its influence experimentally. This is facilitated by the strong dependence on the particle concentration: Scattering of n-th order varies with the n-th power of the concentration. One can therefore reduce the disturbing scattered light drastically by reducing the particle concentration.

5.4 The Path of the Scattered Light

The function of the optical elements in the path of the scattered light is to deliver an undistorted picture of the scattered light field on the observation screen, as was discussed in connection with Fig. 5.3 in Sect. 5.2. There are three aspects which are of special importance:

- The lens L_2 (see Fig. 5.3) must image the focal plane of lens L_1 exactly on the plane of the observation screen (see Sect. 5.2). The distance between the scattering volume and lens L_1 is not of importance. It simply limits the spherical angle within which scattered light is accepted. The larger the distance the smaller the area on the observation screen where scattered light can be observed. This means that only the distances between lens L_1 and lens L_2, and Lens L_2 and the observation screen, must be carefully controlled.
- The laser beam must be imaged exactly on the plane of the diaphragm. Otherwise the region from which scattered light reaches the screen is not well defined and we observed that the speckle motion velocity is decreased.
- If there is a solid surface, (e.g. glass plate or lens), close to the scattering volume, (a few centimetres or less), which the scattered light must pass through, then even small particles on that surface or scratches can lead to severe disturbances of the pattern motion on the screen. This is due to the fact that the coherence area is very small near the scattering volume. It is therefore important to keep such surfaces absolutely clean.

It is generally preferable to have few solid surfaces, which also means optical elements, in the path of the scattered light, since all those surfaces reflect a certain amount of light because of the discontinuity of the refractive index. The reflected light is generally a source of stray light (see next chapter). If mirrors are used one should always prefer surface coated ones.

5.5 Stray Light

We speak of stray light if light, other than the one intended, reaches the detector. In our case two types of stray light may be discriminated: Coherent and incoherent. Incoherent stray light does not interfere with the ‘ordinary’ scattered light and leads only to a background signal. Since all light detectors deliver a signal even in absolute darkness, the so called ‘dark current’ , one may consider the action of the incoherent stray light as an increase of the dark current. This, however, is only true if the temporal fluctuation of the stray light can be neglected as compared to that of the ordinary scattered light.

Coherent stray light actually spoils the speckle pattern in the detector plane, and its influence on the measured speckle motion velocity can be enormous, even at fairly low intensities. It is therefore recommended to eliminate this part of the stray light almost entirely.

There are essentially two sources of incoherent stray light. Firstly, the light in the laboratory, which does not originate from the illuminating laser. Secondly, that light from the laser which has a path length to the detector that is shorter or longer than that of the ordinary scattered light by more than the coherence length of the laser light. (The coherence length of light of a Helium-Neon laser is, for instance, usually 30 cm.)

Sources of coherent stray light could be:

- Successive reflections at optical surfaces, as for instance at lenses and glass plates
- Diffraction and scattering at small inhomogeneities in the beam paths, as for instance at scratches and dust
- Light scattering at particles which are adsorbed at the wall of the flow vessel
- Forward scattering and reflections in the flow vessel.

The last item should be considered more closely: If the diameter of the particles is very small compared to the wavelength of the laser light, the scattered light field is rather isotropic. With increasing diameter, the forward scattering becomes more and more prominent. This may cause difficulties especially for measurements near the wall of the flow vessel and for set-ups where light with small scattering angles should be analysed. It is therefore advisable to choose a scattering geometry where the light scattered in forward direction is not, by reflections at the wall, directed towards the detector.

The last but one item also leads to difficulties for measurements near the walls of the flow vessel, because scattering particles adsorbed at the wall can then contribute to the measured scattered light. The influence of this effect can best be assessed if one successively moves the scattering volume towards the wall. Generally, the wall gradient increases by approaching the wall until it becomes constant very close to the wall itself (c.f. Chap. 5.9.2). The wall gradient is the derivative of the velocity component parallel to the mean flow direction in the direction perpendicular to the wall. If the just mentioned effect is significant, one will notice an increase of the wall gradient followed by a decrease, as soon as the adsorbed particles enter the scattering volume. One can avoid this effect by repeatedly cleaning absorbed particles off the wall. A decrease in the scattering volume size is generally helpful to avoid troubles for measurements near the wall.

5.6 Test of the Set-Up

The optical set-up of any of the measurement methods described below requires a very careful alignment. Otherwise systematic errors in the measured gradients can easily arise, which usually lead to values that are too small. It is therefore advisable to use a light scattering object with a definite velocity gradient in order to test the set-up. Well suited are light scattering plates which rotate with a constant velocity. Then the velocity gradient is exactly known. It is important that the inhomogeneities of this plate are small compared to the diameter of the scattering volume, which is really a scattering area in this case. Otherwise the speckle pattern does not result from many independent scattering particles and one can observe regular patterns, which also may lead to errors in the determined gradient values.

Due to the almost perfect two-dimensionality of the scattering object, the q-volume of coherence is pencil-like. After what we explained above, it is then necessary to have the long axis of this volume parallel to $\boldsymbol{k}_o$, i.e. to the direction of the observed light.

A check of the quality of the set-up consists in measuring the gradient for different locations of the illuminating light beam with respect to the rotation axis. The set-up is only perfect if the determined values do not depend on that position. If, for instance, the illuminating light wave is not a plane wave at the scattering plate, the velocity of the speckle pattern will change with position. This will also happen if the speckle pattern is not observed exactly in the far field.

A disadvantage of this method is the fact that the speckle pattern will be periodic in time. For a method, which is based on temporal averaging, the effective period of this averaging is the time the plate needs for one revolution. Therefore, only a certain amount of different speckles travel past a point in the observation plane, and that might be insufficient for a reliable temporal average.

Preferable to a scattering plate is therefore a rotating cylindrical cuvette which contains a fluid with light scattering particles. If the rotation velocity is constant for some time, the inner fluid will of course rotate with the same velocity. The diffusing particles within the cuvette will prevent the speckle pattern from being periodic in time. A further advantage of this device is that the scattering volume will be the same as in the actual experiment.

5.7 Measurement of Quantities Varying Randomly in Time and Space

We assume that all stochastic quantities, which we deal with in this book, are ergodic. For that reason we do not distinguish between different types of averaging and denote an average value of a quantity A by $\langle A \rangle$, even if a temporal average is meant.

Time dependent quantities will usually not be measured in a continuous way, but rather as a series of data. Hence, one determines a series of values which represent temporal averages over a time interval short compared to the time scale on which the quantity varies. Usually, these short time intervals are chosen in such a way that they have a fixed length - the so called sample time - and a fixed distance between them - the so called dead time. If possible, the dead time can be negligibly short. With the resulting series of values, all interesting quantities can be calculated. Integrals, for instance, will then automatically be represented by sums, which is advantageous for calculations done by computers. Similarly, space dependent quantities (usually one or two dimensional) are represented by series of values. The areas over which these values are integrated are usually called pixels. Again, provision must be made so that size and distance of adjacent pixels is much smaller than the respective coherence length.

It can be stated without exaggeration that any quantity measured in an actual experiment suffers from noise. That means that successive measurements of the same quantity will generally lead to a set of different values. (If the measured quantity is a discrete one, as for instance the number of some objects, there are, of course, also exact measurement values possible) It is usual practice to mentally split a measured value S into a 'true' value S_t and a part S_s, which represents the stochastic behaviour. Therefore, the latter is assumed to be different for any single measurement, whereas the former is just a constant. Furthermore, the average of S_s is assumed to be zero:

$$\langle S_s \rangle = 0. \tag{5.11}$$

Thus, we have for an arbitrary measured value

$$S = S_t + S_s \tag{5.12}$$

and on averaging we get

$$\langle S \rangle = S_t \tag{5.13}$$

If one has a large number of measured values, one evidently has

$$\langle S \rangle_N \approx \frac{1}{N} \sum_{i=1}^{N} (S_{si} + S_t) \approx 0 + \frac{1}{N} S_t \cdot N \tag{5.14}$$

which means

$$\langle S \rangle_N \approx S_t \tag{5.15}$$

The approximation used to describe the connection of the two quantities of the last equation can be quantified by use of a function $H(N, \Delta)$ as follows:

If one calculates the estimated mean value $\langle S \rangle_N$ from a set of N measured values and specifies an interval size Δ, then $H(N, \Delta)$ denotes the probability that the true value S_t is within the interval $[\langle S \rangle_N - \Delta, \langle S \rangle_N + \Delta]$.

$H(N, \Delta)$ depends on the probability distribution of the fluctuating part S_s of the quantity under consideration. A general property of it is that it approaches the value 1 with increasing N and/or Δ, which is not difficult to understand. $H(N, \Delta)$ is known for the case of Gaussian distributions of S_s, for other distribution functions it can usually be calculated only numerically.

To have a certain confidence in an experimentally determined mean value $\langle S\rangle_N$, one usually specifies a value of $H(N,\Delta)$, for instance 0.68. That means one wants a probability of 0.68 that the true mean value lies in the interval $[\langle S\rangle_N - \Delta, \langle S\rangle_N + \Delta]$. If then, for example, the number N of measured values is fixed, it is possible by knowledge of the function $H(N,\Delta)$ to calculate the corresponding Δ, and the result is given in the form $\langle S\rangle_N \pm \Delta$. Sometimes it may also be preferable to first choose Δ and use it to determine the number of measurement values required.

The procedure just sketched is based on the assumption that the measured values are independent. This, however, is often not the case. If, for instance, the mean intensity of the scattered light should be measured with the procedure described in Sect. 3.2, one obtains a series of values for short time intervals. That means that successive values are very much correlated if the time interval between them is shorter than the correlation time of the light intensity. This correlation can be viewed as an effective reduction of the number of measured values.

From the experimental point of view, the best procedure to cope with these problems is to install a real-time measurement method. It is then possible to observe the temporal evolution of the mean value and decide by inspection whether the remaining fluctuations are small enough and the measurement may be stopped. Obviously, this is not really satisfying from a scientific point of view, but most practical cases require such a procedure. The determination of $H(N,\Delta)$ generally requires a much larger effort than the measurement itself, and therefore does not seem feasible.

Temporal correlation functions from time series can be calculated in an especially easy way if one has an equidistant series. The auto-correlation function of a quantity S is usually written as

$$C_S(j) = \langle S(i) \cdot S(i+j)\rangle \,. \tag{5.16}$$

If the temporal interval of two successively measured values is denoted by τ, then $S(i)$ is the mean value of $S(t)$ averaged over the interval $[i\tau, (i+1)\tau]$ (neglecting the dead time). Provided τ is small compared to the correlation time of the quantity $S(t)$, one then obtains instead of the 'exact' correlation function $C_S(t)$, the time averaged correlation function $C_S(j)$ with the averaging interval τ. Written down explicitly, equation (5.16) reads

$$\langle S(i) \cdot S(i+j)\rangle = \frac{1}{N}\sum_{i=1}^{N} S(i) \cdot S(i+j). \tag{5.17}$$

Since, by assumption, we are always dealing with stationary processes, we can also write

$$\langle S(i) \cdot S(i+j)\rangle = \langle S(0) \cdot S(j)\rangle \,. \tag{5.18}$$

5.8 From the Pattern Velocity to the Velocity Gradient

The calculation of the components of the velocity gradient tensor from a set of measured pattern velocities is, in practice, a bit more complicated than one might infer from what was discussed so far. To begin with, we consider the case of one scattering geometry only, which allows the determination of two perpendicular pattern velocity components in the plane of the observation screen. We split this problem into two separated ones, which will be discussed in detail within this section: First, we are going to explain how the quantity $\dot{\boldsymbol{k}}_{o\perp}$ can be determined from the measured pattern velocity, and secondly, we explain how two particular linear combinations of the velocity gradient tensor are calculated from $\dot{\boldsymbol{k}}_{o\perp}$. The necessary calculations require a good command of matrix calculus, even for the fairly simple examples we are going to show explicitly.

5.8.1 The Connection Between Pattern Velocity and $\dot{\boldsymbol{k}}_o$

In Chap. 4.7.1 we discussed how the velocity of the pattern motion is directly connected to the vector $\dot{\boldsymbol{k}}_o$. We are now going to calculate this connection quantitatively. For a given value of $\dot{\boldsymbol{k}}_o$ this velocity is apparently dependent on the position of the observation screen and the positions and focal lengths of the lenses in the path of the scattered light. In addition, it is also affected by other optical elements as for instance glass plates. The influence of such elements on the pattern velocity will now be considered.

The Simple Optical Set-Up of Fig. 4.2. It is not difficult with the help of Fig. 4.2 and pure geometrical considerations to derive the connection between the pattern velocity $\boldsymbol{v}_S$ and $\dot{\boldsymbol{k}}_o$:

$$\frac{\dot{\boldsymbol{k}}_{o\perp}}{|\boldsymbol{k}_o|} = \frac{\boldsymbol{v}_S}{f}. \tag{5.19}$$

The symbol $\dot{\boldsymbol{k}}_{o\perp}$ in this equation denotes the component of the vector $\dot{\boldsymbol{k}}_o$ in the direction perpendicular to $\boldsymbol{k}_o$, f denotes the focal length of the lens L. We discussed already in the paragraphs following equation (4.71) that only this part of $\dot{\boldsymbol{k}}$ is responsible for pattern motion.

The Optical Set-Up of Fig. 5.3. The set-ups shown schematically in the Figs. 4.2 and 5.3 differ in the final section of the scattered light path, which in the set-up of Fig. 5.3 is enhanced by a diaphragm and a further lens. For the pattern velocity $\boldsymbol{v}_E$ in the plane E we obtain from the last equation the relationship

$$\frac{\dot{\boldsymbol{k}}_{o\perp}}{|\boldsymbol{k}_o|} = \frac{\boldsymbol{v}_E}{f_1}. \tag{5.20}$$

The lens L_2 images the plane E on the observation screen S. This causes a difference of the pattern sizes in the two planes. If the distance between the

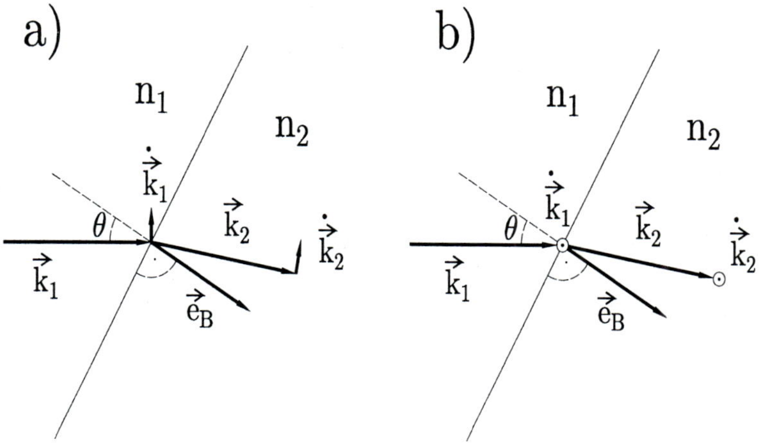

Fig. 5.4. The two special cases of light refraction at a plane boundary, which in a simple way influence the motion of the speckle pattern. Explanation in the text.

plane E and the lens L_2 is denoted by the symbol g_2 and the one from this lens to the observation plane S by b_2 the magnification factor is b_2/g_2. Due to this magnification we obtain the following pattern velocity in the plane S:

$$\boldsymbol{v}_S = \boldsymbol{v}_E \cdot \frac{b_2}{g_2}, \tag{5.21}$$

from which immediately follows

$$\frac{\dot{\boldsymbol{k}}_{o\perp}}{|\boldsymbol{k}_o|} = \frac{\boldsymbol{v}_S}{f_1} \cdot \frac{g_2}{b_2}. \tag{5.22}$$

For the sake of completeness we will quote the famous lens formula , which, with the quantities just introduced, reads

$$\frac{1}{b_2} + \frac{1}{g_2} = \frac{1}{f_2}. \tag{5.23}$$

Refraction at Planes. In general, the scattering volume will be situated in a medium with a refractive index different from that one at the observation screen. The refraction which takes place at the boundary between these two media usually causes a change in the direction of the scattered light. In most cases this boundary lies between the scattering volume and the first lens. It is then *not* important where exactly this plane lies; only its exact orientation with respect to the optical axis influences the behaviour of the pattern motion. We will now only consider this situation.

Even in the above mentioned cases, the problem to be solved is not simple, since the speckle motion depends on the angle between the direction of the scattered light and the boundary plane, and also on the motion direction of the scattered light. Again, we will only consider two special cases, which are represented schematically in Fig. 5.4. (In this figure the index o has been omitted.) These two cases are defined by the directions of the components of the vector $\dot{\boldsymbol{k}}_{o\perp}$, which are measured. In the first case that component is measured, which describes a motion in the plane defined by the direction of $\boldsymbol{k}_o$ and the normal $\boldsymbol{e}_B$ of the refracting plane. In the other case, it is just the component perpendicular to that one. (In the figure, the refracting plane is perpendicular to the plane of the picture). We denote the angle between the two vectors by θ, and the wave vector $\boldsymbol{k}_o$ in the medium i by $\boldsymbol{k}_i$. For the first case, (part a), of Fig. 5.4 we have

$$(\dot{\boldsymbol{k}}_{o\perp})_2 = (\dot{\boldsymbol{k}}_{o\perp})_1 \tag{5.24}$$

and for the second case, (part b), of Fig. 5.4):

$$(\dot{\boldsymbol{k}}_{o\perp})_2 = (\dot{\boldsymbol{k}}_{o\perp})_1 \cdot \frac{n_{12} \cdot \cos(\theta)}{\sqrt{n_{12}^2 - \sin^2(\theta)}}. \tag{5.25}$$

In the last equation we used the notation

$$n_{12} = \frac{n_1}{n_2}. \tag{5.26}$$

One should be aware of the fact that the first equation does *not* mean that the pattern velocity remains unaffected by the refraction. In this case, it is important that the length of the wave vector in the two media is different, which follows from equations (4.13) and (4.15). One generally has

$$\frac{\lambda_1}{\lambda_2} = \frac{n_2}{n_1} = \frac{1}{n_{12}} \tag{5.27}$$

from which follows

$$\frac{|\boldsymbol{k}_1|}{|\boldsymbol{k}_2|} = \frac{n_1}{n_2} = n_{12}. \tag{5.28}$$

The connection of the angular velocities in the two media (see equation 4.73) can then be worked out for the first case to be

$$\frac{(\dot{\boldsymbol{k}}_{o\perp})_2}{|\boldsymbol{k}_2|} = n_{12} \cdot \frac{(\dot{\boldsymbol{k}}_{o\perp})_1}{|\boldsymbol{k}_1|} \tag{5.29}$$

and for the second:

$$\frac{(\dot{\boldsymbol{k}}_{o\perp})_2}{|\boldsymbol{k}_2|} = \frac{(\dot{\boldsymbol{k}}_{o\perp})_1}{|\boldsymbol{k}_1|} \cdot \frac{n_{12}^2 \cdot \cos(\theta)}{\sqrt{n_{12}^2 - \sin^2(\theta)}}. \tag{5.30}$$

The general connection between $(\dot{\boldsymbol{k}}_{o\perp})_1$ and $(\dot{\boldsymbol{k}}_{o\perp})_2$ is rather complicated, especially with respect to the directions. One should therefore restrict the measurements to the components just considered.

It is possible to avoid the necessity to discriminate between these two cases if one takes care of the following items:

- The direction of a light wave which passes through a medium with parallel boundary surfaces, (parallel plate), is not altered. Only a shift of the wave occurs, parallel to the direction of the boundary plane, if the wave does not impinge on the surface at right angles. This means, for instance, that for the transition 'water - glass plate - air' only the transition 'water - air' must be considered.
- If scattered light within a water vessel is to be measured, which passes through the bordering transparent wall not at right angles one can circumvent the complications just discussed. Attach a water filled cuvette, of such a shape that the light leaves the cuvette at right angles, to the outside of the wall. In that case one has $\theta = 0$ and the two equations (5.24) and (5.25) give the same result. Also, the two equations (5.24) and (5.29) are then identical, which is true for arbitrary speckle motion directions.

Reflection at a Mirror. In cases where the scattered light in an oblique direction has to be analysed it may be advantageous to guide it by means of one or more mirrors in a more suitable direction. So one will usually prefer the detection channel - i.e. that part of an experimental set-up where the scattered light is analysed - with its lenses, pinholes, and detectors to be mounted on an optical bench placed on a table. This requires that the light to be analysed is reflected in a horizontal direction. It must then be taken into account that the direction of the pattern motion is influenced by the reflection as well, and it is not easy to obtain the altered pattern motion direction.

Mathematically, the action of a mirror on the direction of a light wave can be described by a certain class of orthogonal matrices with a determinant -1. So for instance a mirror with a plane parallel to the x-y-plane is described by the following matrix

$$\hat{O}^m = \begin{pmatrix} 1 & 0 & 0 \\ 0 & 1 & 0 \\ 0 & 0 & -1 \end{pmatrix} \tag{5.31}$$

One easily convinces oneself that a vector pointing in the z-direction changes sign if multiplied by this matrix, whereas vectors in the x-y-plane, i.e. those with a vanishing z-component, remain unaltered.

A mirror oriented in an arbitrary direction can always be obtained from one with special orientation by rotating it properly. Consequently, the matrix describing such a rotated mirror is obtained from the matrix corresponding to the special mirror by properly rotating it according to equation (3.91) in Chap. 3.6. For that purpose, the necessary rotation must be known, i. e. its rotation axis and rotation angle.

In practice, however, one usually adjusts a mirror experimentally in such a way that a certain light wave travels - after reflection - in the desired direction. For the construction of the corresponding reflection matrix one has to know the directions of the light wave prior and after the reflection. Let us denote these directions by the corresponding unit-vectors: $\boldsymbol{e}_o$ denotes the direction of the light wave impinging on the mirror and $\boldsymbol{e}_r$ gives the wave direction after reflection. The mirror is uniquely determined by the vector $\boldsymbol{e}_m$ normal to the reflection plane. The basic property of mirrors is that the angles formed of the mentioned two waves form like angles with that normal of the mirror plane. From that property immediately follows

$$\boldsymbol{e}_m = \frac{\boldsymbol{e}_r - \boldsymbol{e}_o}{|\boldsymbol{e}_r - \boldsymbol{e}_o|} \tag{5.32}$$

It is not very difficult to work out the matrix $\hat{O}^r$ which has the desired property to transform a vector parallel to $\boldsymbol{e}_m$ into its negative but leaving vectors perpendicular to $\boldsymbol{e}_m$ unchanged. If one writes $\boldsymbol{e}_m = (a, b, c)$ one obtains the following form of the desired mirror matrix:

$$\hat{O}^m = \begin{pmatrix} 1-2a^2 & -2ab & -2ac \\ -2ab & 1-2b^2 & -2bc \\ -2ac & -2bc & 1-2c^2 \end{pmatrix} \tag{5.33}$$

With this reflection matrix one can now calculate in what way the pattern motion direction is changed after the scattered light wave has passed the corresponding mirror. It is worth noting that any reflection matrix has the following properties:

- its trace is 1
- it is symmetric
- its determinant is -1

Especially the first two properties can be used as a first check of a calculated reflection matrix.

Examples of the application of these considerations are given in Chap. 5.9.

5.8.2 The Connection Between $\hat{\Gamma}$ and $\dot{\boldsymbol{k}}_o$ – an Example

In this chapter we are going to demonstrate, by means of an example, the procedure to obtain the quantity $\hat{\Gamma}$ from equation (4.72). The example refers to the set-up shown in Fig. 8.2. This figure shows a set-up where the light scattering takes place in a flow channel, the illuminating laser beam is drawn as a thick line in direction x'. The light scattered in direction y' is measured. Consequently, the directions of the two wave vectors $\boldsymbol{k}_i$ and $\boldsymbol{k}_o$ are known. The wavelength of the light source is assumed to be known as well, although it is only a dummy variable. It will not enter the final result. In the mentioned figure, two co-ordinate systems are drawn, called S and S′. The system S′ is most suitable for the description of the considered light waves ('the system of the light'), whereas the system S is best adapted to the flow ('the system of the flow'). The two systems are rotated with respect to each other by an angle of 45 degrees about their common z-axis. To begin with, we will perform the necessary calculations in the system S′.

Calculation in the System of the Light. The unit-vectors, which denote the directions of the illuminating and the detected light, are given by

$$\begin{aligned} \boldsymbol{e}_i{}' &= (1,0,0) \\ \boldsymbol{e}_o{}' &= (0,1,0), \end{aligned} \tag{5.34}$$

respectively. The primed quantities in this section always indicate that they are expressed in reference to the co-ordinate system S′. We will assume that the two orthogonal velocity components of the stray light pattern, which are to be measured, are the ones in x′- and in z′-direction, respectively. Then we immediately have the two vectors

$$\begin{aligned} \dot{\boldsymbol{k}}_1' &= (\dot{k}_{ox'},0,0) \\ \dot{\boldsymbol{k}}_2' &= (0,0,\dot{k}_{oz'}). \end{aligned} \tag{5.35}$$

Without question, these two velocity components can be measured, because they are orthogonal to $\boldsymbol{k}_o'$. According to equation (4.72) we then get

$$\frac{\dot{\boldsymbol{k}}_o'}{|\boldsymbol{k}_o|} = \Gamma'^T \cdot (\boldsymbol{e}_i{}' - \boldsymbol{e}_o{}') \tag{5.36}$$

$$\begin{aligned} \frac{\dot{\boldsymbol{k}}_o'}{|\boldsymbol{k}_o|} &= \begin{pmatrix} \Gamma_{x'x'} & \Gamma_{y'x'} & \Gamma_{z'x'} \\ \Gamma_{x'y'} & \Gamma_{y'y'} & \Gamma_{z'y'} \\ \Gamma_{x'z'} & \Gamma_{y'z'} & \Gamma_{z'z'} \end{pmatrix} \cdot \begin{pmatrix} 1 \\ -1 \\ 0 \end{pmatrix} \\ &= \begin{pmatrix} \Gamma_{x'x'} - \Gamma_{y'x'} \\ \Gamma_{x'y'} - \Gamma_{y'y'} \\ \Gamma_{x'z'} - \Gamma_{y'z'} \end{pmatrix}. \end{aligned} \tag{5.37}$$

By inspection of this equation the two desired components can now be obtained:

$$\begin{aligned} \frac{\dot{k}_{ox'}}{|\boldsymbol{k}_o|} &= (\Gamma_{x'x'} - \Gamma_{y'x'}) \\ \frac{\dot{k}_{oz'}}{|\boldsymbol{k}_o|} &= (\Gamma_{x'z'} - \Gamma_{y'z'})\,. \end{aligned} \tag{5.38}$$

In this equation, the components of the velocity gradient tensor are expressed in a rather unsuitable manner. It would be much more convenient to have them expressed in the system S. Since the system S' can be transformed to the system S by a rotation around the z-axis about 45°, we have to build the rotation matrix for this transformation (Note that z- and z′-axes are identical). According to equation (3.75) we then obtain

$$\hat{O} = \begin{pmatrix} \cos(45^o) & \sin(45^o) & 0 \\ -\sin(45^o) & \cos(45^o) & 0 \\ 0 & 0 & 1 \end{pmatrix} = \frac{1}{\sqrt{2}} \begin{pmatrix} 1 & 1 & 0 \\ -1 & 1 & 0 \\ 0 & 0 & \sqrt{2} \end{pmatrix}. \tag{5.39}$$

The transformed gradient tensor can now be calculated according to equation (3.90):

$$\hat{\Gamma}' = \hat{O} \cdot \hat{\Gamma} \cdot \hat{O}^T. \tag{5.40}$$

Performing the multiplications on the right hand side of this equation yields

$$\begin{pmatrix} \Gamma_{x'x'} & \Gamma_{y'x'} & \Gamma_{z'x'} \\ \Gamma_{x'y'} & \Gamma_{y'y'} & \Gamma_{z'y'} \\ \Gamma_{x'z'} & \Gamma_{y'z'} & \Gamma_{z'z'} \end{pmatrix} = \tag{5.41}$$

$$\frac{1}{2} \begin{pmatrix} \Gamma_{xx} + \Gamma_{xy} + \Gamma_{yx} + \Gamma_{yy} & -\Gamma_{xx} - \Gamma_{xy} + \Gamma_{yx} + \Gamma_{yy} & \sqrt{2}(\Gamma_{zx} + \Gamma_{zy}) \\ -\Gamma_{xx} + \Gamma_{xy} - \Gamma_{yx} + \Gamma_{yy} & \Gamma_{xx} - \Gamma_{xy} - \Gamma_{yx} + \Gamma_{yy} & \sqrt{2}(-\Gamma_{zx} + \Gamma_{zy}) \\ \sqrt{2}(\Gamma_{xz} + \Gamma_{yz}) & \sqrt{2}(-\Gamma_{xz} + \Gamma_{yz}) & 2\Gamma_{zz} \end{pmatrix}.$$

It is now fairly easy to express the right hand sides of equation (5.38) by the components of $\hat{\Gamma}$ in the system S:

$$\begin{aligned} (\Gamma_{x'x'} - \Gamma_{y'x'}) &= \frac{1}{2}(\Gamma_{xx} + \Gamma_{xy} + \Gamma_{yx} + \Gamma_{yy}) - \frac{1}{2}(-\Gamma_{xx} - \Gamma_{xy} + \Gamma_{yx} + \Gamma_{yy}) \\ &= (\Gamma_{xx} + \Gamma_{xy}) \\ (\Gamma_{x'z'} - \Gamma_{y'z'}) &= \frac{1}{2}(\Gamma_{xz} + \Gamma_{yz}) - \frac{1}{2}(-\Gamma_{xz} + \Gamma_{yz}) \\ &= \Gamma_{xz}. \end{aligned} \tag{5.42}$$

We finally see that a set-up with a scattering geometry shown in Fig. 8.2 allows for the measurement of the component Γ_{xz} and of the linear combination $\Gamma_{xx} + \Gamma_{xy}$. If other components of the pattern velocity are measured one obtains linear combinations of these two components.

Calculation in the System of the Flow. The unit-vectors introduced in the last section (see equation (5.34) have in the system S the components

$$\begin{aligned} \boldsymbol{e}_i &= \frac{1}{\sqrt{2}}(1,1,0) \\ \boldsymbol{e}_o &= \frac{1}{\sqrt{2}}(-1,1,0). \end{aligned} \tag{5.43}$$

Consequently, we get

$$\boldsymbol{e}_i - \boldsymbol{e}_o = (\sqrt{2}, 0, 0). \tag{5.44}$$

The two orthogonal components of the pattern velocity considered in the last section correspond to the directions of the following unit-vectors:

$$\begin{aligned} \boldsymbol{e}_{P1} &= \frac{1}{\sqrt{2}}(1,1,0) \\ \boldsymbol{e}_{P2} &= (0,0,1). \end{aligned} \tag{5.45}$$

The connection between the pattern velocity and the corresponding gradient component can now be obtained if one of the general equations (4.72) or (5.36) is multiplied with the respective unit-vector. This will now be done for the component 1:

$$\begin{aligned} \frac{\dot{\boldsymbol{k}}_o \cdot \boldsymbol{e}_{P1}}{|\boldsymbol{k}_o|} &= \left(\hat{\Gamma}^T \cdot (\boldsymbol{e}_i - \boldsymbol{e}_o)\right)\boldsymbol{e}_{P1} \\ &= \left(\hat{\Gamma}^T \cdot (\sqrt{2},0,0)\right) \cdot \boldsymbol{e}_{P1}. \end{aligned} \tag{5.46}$$

From this equation follows

$$\begin{aligned} \frac{\dot{k}_{o1}}{|\boldsymbol{k}_o|} &= \sqrt{2}\begin{pmatrix} \Gamma_{xx} \\ \Gamma_{xy} \\ \Gamma_{xz} \end{pmatrix} \cdot \frac{1}{\sqrt{2}}(1,1,0) \\ &= \Gamma_{xx} + \Gamma_{xy}. \end{aligned} \tag{5.47}$$

For the second component, the multiplication with $\boldsymbol{e}_{P2}$ is especially simple and we obtain, as before:

$$\frac{\dot{k}_{o2}}{|\boldsymbol{k}_o|} = \Gamma_{xz}. \tag{5.48}$$

Which of the two ways described here one prefers is, in the end, a matter of taste. Generally, one wants to have the gradient components in the system determined by the flow, and for that reason the calculation performed in the second section seems to be more suitable. It has the advantage that no transformation of the gradient tensor has to be performed. One has to know, however, the unit-vectors in the directions of the illuminating light beam, the detected wave component, and the considered pattern velocity components, expressed in the system of the flow. This is, for arbitrary directions of these vectors, no simple task, but cannot be circumvented.

5.9 Measurement of Special Components of the Velocity Gradient Tensor

5.9.1 Measurement of Definite Cartesian Components

With the techniques described here there is a significant difference between the measurement of on-diagonal or off-diagonal elements of the velocity gradient tensor, independent of the chosen co-ordinate system. There always exists a scattering geometry to measure exactly one definite off-diagonal element

$\frac{\partial v_i}{\partial x_j}$, just choose the scattering plane parallel to the direction i and align $\boldsymbol{k}_i$ and $\boldsymbol{k}_o$ within that plane such that the resulting $\boldsymbol{q}$ points in the direction i. Then measure the velocity of the pattern in direction j.

For the measurement of on-diagonal elements the vectors $\boldsymbol{q}$ and $\boldsymbol{k}_o$ must be orthogonal. It is easy to convince oneself that this is not possible. So it turns out that on-diagonal elements can only be measured as linear combinations containing at least one off-diagonal element. Consequently, at least two scattering set-ups are required to measure one on-diagonal element.

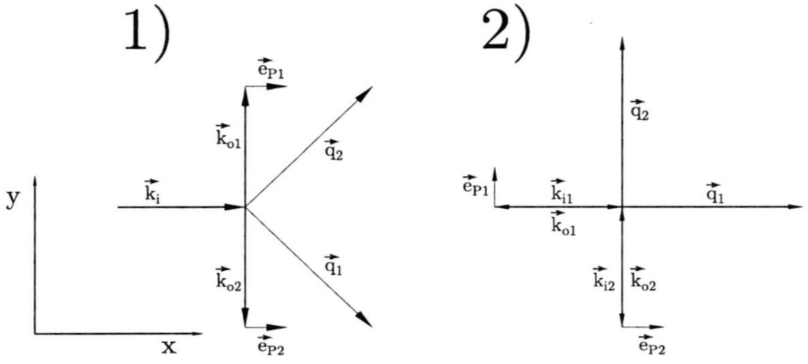

Fig. 5.5. Scattering geometries for measurements of 1): the on-diagonal element $\frac{\partial v_x}{\partial x}$ and 2) the vorticity component $\frac{\partial v_x}{\partial y} - \frac{\partial v_y}{\partial x}$. Part 1): All vectors are in the x-y plane. Part 2) The vectors $\boldsymbol{k}_i$ and $\boldsymbol{k}_o$ form angles of 45° with the x-y plane in such a way that the vectors $\boldsymbol{q}_1$ and $\boldsymbol{q}_2$ lie in the x-y plane. The vectors $\boldsymbol{e}_{Pi}$ point in the direction in which the pattern velocity has to be measured. Further explanation in the text.

Figure 5.5 part 1) shows schematically the scattering geometry for an optical set-up which allows such a measurement. There is one illuminating light wave with wave vector $\boldsymbol{k}_i$ and two components of the scattered light with the wave vectors $\boldsymbol{k}_{o1}$ and $\boldsymbol{k}_{o2}$ are analysed. The components of the pattern velocities are those parallel to the vectors $\boldsymbol{e}_{P1}$ and $\boldsymbol{e}_{P2}$, respectively. According to equation (5.46) with the scattering channel 1 the linear combination $\frac{\partial v_x}{\partial x} - \frac{\partial v_y}{\partial x}$ is measured, with the channel 2 it is $\frac{\partial v_x}{\partial x} + \frac{\partial v_y}{\partial x}$.

The determination of a vorticity component requires the measurement of two components of the velocity gradient tensor. Figure 5.5 part 2) shows schematically how the geometry of a possible set-up could look like. Actually two scattering experiments are performed, with the one denoted by the superscript 1 the component $\frac{\partial v_x}{\partial y}$ and with the other one $\frac{\partial v_y}{\partial x}$ is measured. Of course, with these two components also the shear component $\frac{\partial v_x}{\partial y} + \frac{\partial v_y}{\partial x}$ can be determined. A complication in this set-up is that light from the laser beam 1 reaches the detectors of the set-up 2 and vice versa. That can be prevented by switching the two illuminating laser beams and gating the signals of the detectors respectively.

It will not always be possible to build the necessary optical set-up for a special gradient component due to restrictions caused by the flow geometry (see for instance next subsection). In such cases a solution to this problem might be the following one: Instead of using **one** illuminating light wave one uses two and also two detectors. It is explained and demonstrated in Chap. 10 how it is possible to measure with this technique a single off-diagonal element of $\hat{\Gamma}$ with a scattering geometry which in the simple version - one illuminating light beam - would deliver a linear combination with an on-diagonal element.

5.9.2 Measurements of Wall Gradients

Next to a solid boundary the fluid velocity can have two non-vanishing orthogonal components only, since the component normal to the wall must vanish. Furthermore, these components can only change in the direction normal to the wall. So we are left with just two independent components of the average velocity gradient at the wall. If one chooses a co-ordinate system with one axis in the direction of the average flow and the other normal to the wall face, then there will be only one average gradient component, which is called the wall gradient. There is always a certain layer of the fluid near the wall where the velocity gradient can be viewed as constant, the so-called viscous sublayer. The thickness of this layer depends on properties of the flow and the fluid. If one wants to measure the wall gradient one must therefore use a scattering volume not larger than the width of this layer.

Because of mentioned restriction at the wall there is a great liberty as far as the scattering geometry is concerned. Two scattering geometries will now be discussed in detail, which are depicted schematically in Fig. 5.6.

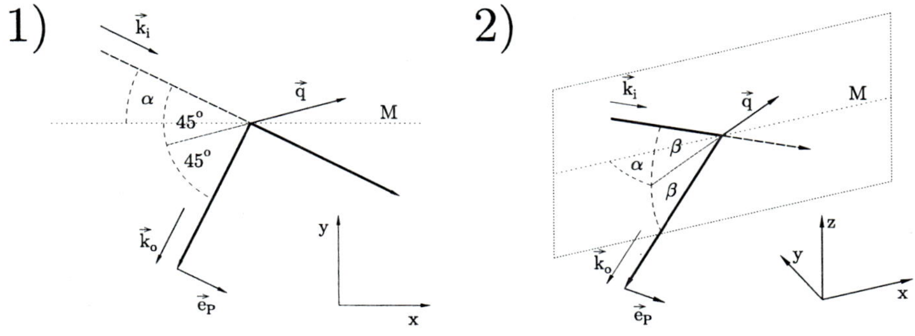

Fig. 5.6. Scattering geometries to measure wall gradients. The illuminating light beam within the fluid is represented by a dotted line. Explanation in the text.

Geometry 1 is convenient for flows where the main velocity is in the horizontal plane. The wall gradient is then the change of the velocity also in a horizontal direction and consequently all light paths can stay in the horizontal. This makes the control and determination of the necessary angles fairly simple. Unfortunately, the illuminating light must traverse the flow, which generally reduces the intensity of the scattered light and furthermore, two optical windows are required at opposite sides of the flow tube. In the second geometry, the light paths are very short, but they are more difficult to control, and the optical set-up will be a lot more complicated.

Let us now derive the connections between $\hat{\Gamma}$ and $\dot{\boldsymbol{k}}_{o\perp}$ for these two scattering geometries. In both cases the velocity gradient tensor in the viscous sublayer and for the given co-ordinate system has the following form:

$$\Gamma = \begin{pmatrix} 0 & \Gamma_{xy} & 0 \\ 0 & 0 & 0 \\ 0 & \Gamma_{zy} & 0 \end{pmatrix}. \tag{5.49}$$

Scattering geometry 1. The important k-vectors can easily be obtained by inspection of part 1 of Fig. 5.6:

$$\begin{aligned} \boldsymbol{k}_i &= k \begin{pmatrix} \cos(\alpha) \\ -\sin(\alpha) \\ 0 \end{pmatrix} = k \cdot \boldsymbol{e}_i \\ \boldsymbol{k}_o &= k \begin{pmatrix} -\sin(\alpha) \\ -\cos(\alpha) \\ 0 \end{pmatrix} = k \cdot \boldsymbol{e}_o \\ \boldsymbol{q} &= k \begin{pmatrix} \cos(\alpha) + \sin(\alpha) \\ \cos(\alpha) - \sin(\alpha) \\ 0 \end{pmatrix} = k \cdot (\boldsymbol{e}_i - \boldsymbol{e}_o). \end{aligned} \tag{5.50}$$

Inserting these vectors in equation 4.70 yields

$$\dot{\boldsymbol{k}}_o = \begin{pmatrix} 0 & 0 & 0 \\ \Gamma_{xy} & 0 & \Gamma_{zy} \\ 0 & 0 & 0 \end{pmatrix} \cdot k \begin{pmatrix} \cos(\alpha) + \sin(\alpha) \\ \cos(\alpha) - \sin(\alpha) \\ 0 \end{pmatrix} = k\Gamma_{xy} \cdot \begin{pmatrix} 0 \\ \cos(\alpha) + \sin(\alpha) \\ 0 \end{pmatrix}. \tag{5.51}$$

The observed speckle motion is given by the projection of this vector onto the observation plane, i.e. on the plane to which the vector $\boldsymbol{k}_o$ is normal. This projection can be performed by subtracting from $\dot{\boldsymbol{k}}_o$ that part which is parallel to $\boldsymbol{k}_o$:

$$\dot{\boldsymbol{k}}_{o\perp} = \dot{\boldsymbol{k}}_o - \left(\dot{\boldsymbol{k}}_o \cdot \boldsymbol{e}_o\right) \boldsymbol{e}_o. \tag{5.52}$$

The simple calculation yields

$$\dot{\boldsymbol{k}}_{o\perp} = -k \cdot \Gamma_{xy} \left(\sin^2(\alpha) + \cos(\alpha)\sin(\alpha)\right) \cdot \begin{pmatrix} \cos(\alpha) \\ -\sin(\alpha) \\ 0 \end{pmatrix} \tag{5.53}$$

The pattern moves in the x-y-plane, exactly opposite to the direction of the illuminating light beam. The rate of change of the respective component of the k-vector is:

$$\dot{k}_{o\perp} = k \cdot \Gamma_{xy} \cdot (\sin^2(\alpha) + \cos(\alpha)\sin(\alpha)) \tag{5.54}$$

Let us now assume that a mirror is placed in the path of the scattered light in such a way that a wave with wave vector $\boldsymbol{k}_o$ is transformed to a wave with wave vector $k(0, -1, 0)$. From these two vectors the normal of the mirror plane can be obtained according to equation (5.32):

$$\begin{aligned} \boldsymbol{e}_m &= \frac{1}{\sqrt{\sin^2(\alpha) + (\cos(\alpha) - 1)^2}} \begin{pmatrix} \sin(\alpha) \\ \cos(\alpha) - 1 \\ 0 \end{pmatrix} \\ &= \frac{1}{\sqrt{2(1 - \cos(\alpha))}} \begin{pmatrix} \sin(\alpha) \\ \cos(\alpha) - 1 \\ 0 \end{pmatrix} \end{aligned} \tag{5.55}$$

The matrix describing the mirror action can now be calculated on the basis of equation (5.33):

$$\hat{O}^m = \begin{pmatrix} 1 - \frac{2\sin^2(\alpha)}{2(1-\cos(\alpha))} & -\frac{2(\sin^2(\alpha)-\sin(\alpha)\cos(\alpha)}{2(1-\cos(\alpha))} & 0 \\ -\frac{2(\sin^2(\alpha)-\sin(\alpha)\cos(\alpha)}{2(1-\cos(\alpha))} & 1 - \frac{2(\cos(\alpha)-1)^2}{2(1-\cos(\alpha))} & 0 \\ 0 & 0 & 1 \end{pmatrix} \tag{5.56}$$

With the help of relations between the trigonometric functions this matrix can be simplified with the result

$$\hat{O}^m = \begin{pmatrix} -\cos(\alpha) & \sin(\alpha) & 0 \\ \sin(\alpha) & \cos(\alpha) & 0 \\ 0 & 0 & 1 \end{pmatrix} \tag{5.57}$$

It is not difficult to verify the relation

$$\hat{O}^m \cdot \boldsymbol{k}_o = k \begin{pmatrix} 0 \\ -1 \\ 0 \end{pmatrix} = \boldsymbol{k}_{om} \tag{5.58}$$

where we defined the vector $\boldsymbol{k}_{om}$. The pattern motion is now given by

$$\hat{O}^m \cdot \dot{\boldsymbol{k}}_o = k\Gamma_{xy} \begin{pmatrix} \cos(\alpha)\sin(\alpha) + \sin^2(\alpha) \\ \cos^2(\alpha) + \cos(\alpha)\sin(\alpha) \\ 0 \end{pmatrix} = \dot{\boldsymbol{k}}_{om} \tag{5.59}$$

The component of $\dot{\boldsymbol{k}}_{om}$ perpendicular to the direction of $\boldsymbol{k}_{om}$ is apparently the vector

$$\dot{\boldsymbol{k}}_{om\perp} = k\Gamma_{xy}\begin{pmatrix}\cos(\alpha)\sin(\alpha)+\sin^2(\alpha)\\ 0\\ 0\end{pmatrix} \tag{5.60}$$

which has of course the same absolute value as the vector $\dot{\boldsymbol{k}}_{o\perp}$ in equation (5.54).

Scattering geometry 2. The situation here is a bit more difficult. We start with an intermediate scattering geometry where the vectors $\boldsymbol{k}_i$ and $\boldsymbol{k}_o$ lie in the x-z-plane. The final geometry is then obtained by this one through a rotation by the angle α around the z-axis. As we will see it is not obvious in this case in which direction the speckles move. The vectors in the intermediate geometry are:

$$\boldsymbol{k}_i^I = k\begin{pmatrix}\cos(\beta)\\ 0\\ -\sin(\beta)\end{pmatrix} \tag{5.61}$$

$$\boldsymbol{k}_o^I = k\begin{pmatrix}-\cos(\beta)\\ 0\\ -\sin(\beta)\end{pmatrix}.$$

These vectors must now be rotated around the positive z-axis, the corresponding rotation matrix is given by equation (3.76):

$$\hat{O}_z^v(\alpha) = \begin{pmatrix}\cos(\alpha) & -\sin(\alpha) & 0\\ \sin(\alpha) & \cos(\alpha) & 0\\ 0 & 0 & 1\end{pmatrix}. \tag{5.62}$$

The multiplication of this matrix with the vectors in equation (5.61) gives

$$\boldsymbol{k}_i = k\begin{pmatrix}\cos(\alpha)\cdot\cos(\beta)\\ \sin(\alpha)\cdot\cos(\beta)\\ -\sin(\beta)\end{pmatrix} = k\boldsymbol{e}_i \tag{5.63}$$

$$\boldsymbol{k}_o = k\begin{pmatrix}-\cos(\alpha)\cdot\cos(\beta)\\ -\sin(\alpha)\cdot\cos(\beta)\\ -\sin(\beta)\end{pmatrix} = k\boldsymbol{e}_o$$

$$\boldsymbol{q} = k\begin{pmatrix}2\cos(\alpha)\cdot\cos(\beta)\\ 2\sin(\alpha)\cdot\cos(\beta)\\ 0\end{pmatrix} = k(\boldsymbol{e}_i - \boldsymbol{e}_o).$$

The speckle velocity is now given by

$$\dot{\boldsymbol{k}}_o = k\left(\Gamma^T\cdot(\boldsymbol{e}_i - \boldsymbol{e}_o)\right) \tag{5.64}$$

$$= k\begin{pmatrix}0 & 0 & 0\\ \Gamma_{xy} & 0 & \Gamma_{zy}\\ 0 & 0 & 0\end{pmatrix}\begin{pmatrix}2\cos(\alpha)\cos(\beta)\\ 2\sin(\alpha)\cos(\beta)\\ 0\end{pmatrix}$$

$$= 2k\Gamma_{xy}\cdot\cos(\alpha)\cos(\beta). \tag{5.65}$$

Analogous to the procedure outlined in the preceding paragraph, we could again calculate the projection of this vector onto the observation plane according to equation (5.52). However, it turns out that the pattern motion in this case is not in the x-y-plane. The result one obtains is a rather complicated expression of trigonometric functions of α and β. We therefore preferred to calculate the projection of the pattern velocity within the observation plane onto the y-axis. For that purpose we construct a unit-vector which lies in the x-y-plane and which is perpendicular to $\boldsymbol{k}_o$. It is not difficult to verify that the vector $\boldsymbol{e}_P = (-\sin(\alpha), \cos(\alpha), 0)$ has just this property. So we calculate the component of $\dot{\boldsymbol{k}}_o$ in the direction of $\boldsymbol{e}_P$:

$$\dot{k}_P = \dot{\boldsymbol{k}}_o \cdot \boldsymbol{e}_P = 2k\Gamma_{xy}\cos^2(\alpha)\cos^2(\beta). \tag{5.66}$$

It is obvious from an experimental point of view that a measurement of the pattern velocity can much easier be performed if the scattered light travels in the x-y-plane. Let us therefore calculate the pattern velocity after the reflection of the scattered light at a mirror which directs it in the negative y-direction.

The mirror vector $\boldsymbol{e}_m$ is determined by $\boldsymbol{e}_o$ and the vector $(0, -1, 0)$ (see preceding paragraph):

$$\boldsymbol{e}_m = \frac{1}{\sqrt{2(1-\sin(\alpha)\cdot\cos(\beta))}} \begin{pmatrix} \cos(\alpha)\cdot\cos(\beta) \\ \sin(\alpha)\cdot\cos(\beta) - 1 \\ \sin(\beta) \end{pmatrix}. \tag{5.67}$$

Unfortunately, the reflection matrix in this case contains rather complicated expressions:

$$\hat{O}^m = \begin{pmatrix} 1 - 2N^{-1}\cos^2(\alpha)\cos^2(\beta) & \cos(\alpha)\cos(\beta) & N^{-1}\cos(\alpha)\sin(2\beta) \\ \cos(\alpha)\cos(\beta) & \cos(\alpha)\sin(\beta) & \sin(\beta) \\ N^{-1}\cos(\alpha)\sin(2\beta) & \sin(\beta) & 1 - 2N^{-1}\sin^2(\beta) \end{pmatrix} \tag{5.68}$$

where the factor N is given by

$$N = 2\cdot(1 - \sin(\alpha)\cos(\beta)). \tag{5.69}$$

In spite of that rather complex matrix the multiplication of it with the vector $\dot{\boldsymbol{k}}_o$ yields a rather simple result:

$$\hat{O}^m \cdot \dot{\boldsymbol{k}}_o = \dot{\boldsymbol{k}}_{om} = k\Gamma_{xy} \begin{pmatrix} 2\cos^2(\alpha)\cos^2(\beta) \\ \cos^2(\beta)\sin(2\alpha) \\ \cos(\alpha)\sin(2\beta) \end{pmatrix}. \tag{5.70}$$

The projection of this vector on the observation plane, i.e. on the plane perpendicular to $\boldsymbol{k}_{om}$ is

$$\dot{\boldsymbol{k}}_{om\perp} = k\Gamma_{xy} \begin{pmatrix} 2\cos^2(\alpha)\cos^2(\beta) \\ 0 \\ \cos(\alpha)\sin(2\beta) \end{pmatrix}. \tag{5.71}$$

One notes that the pattern doesn't move in the horizontal plane. It is of course possible to determine the direction of the speckle motion. For that purpose we write

$$\dot{\boldsymbol{k}}_{om\perp} = C \cdot \begin{pmatrix} \cos(\phi) \\ 0 \\ \sin(\phi) \end{pmatrix} = \begin{pmatrix} A \\ 0 \\ B \end{pmatrix} \quad (5.72)$$

where ϕ is the unknown angle which the pattern motion direction forms with the x-axis. It is clear from this equation that

$$\tan(\phi) = \frac{B}{A} = \frac{\sin(2\beta)}{2\cos(\alpha)\cos^2(\beta)}. \quad (5.73)$$

The pattern velocity in that direction is given by the absolute value of the vector $\dot{\boldsymbol{k}}_{om\perp}$ which is

$$|\dot{\boldsymbol{k}}_{om\perp}| = k\Gamma_{xy}\sqrt{4\cos^4(\alpha)\cos^4(\beta) + \cos^2(\alpha)\sin^2(2\beta)}. \quad (5.74)$$

5.10 General Notes to Channel Flows

In the following chapters of this book we will show experimental results of measurement methods that are based on the theory outlined in the preceding chapters. Because all of the measurements performed so far were done on channel flows, or flows in rectangular flow tubes, we summarize here the common notations for these types of flows. The Cartesian co-ordinate system is chosen such that its x-axis points in the direction of the flow, the y-axis is normal to the side wall and points from the wall into the flow. The z-axis is chosen as to obtain a right handed co-ordinate system (see also Fig. 8.2). For the components of the velocity in this co-ordinate system the following notation is in use:

$$\boldsymbol{u} = (u_x, u_y, u_z) = (u, v, w). \quad (5.75)$$

The Reynolds number R_e of a channel flow is expressed by the velocity u_s in the middle of the tube or channel, the kinematic viscosity ν_S (see equation (2.6)) and the width b of the channel or tube:

$$R_e = \frac{b \cdot u_s}{\nu_S}. \quad (5.76)$$

For a comparison of flows with different Reynolds numbers, one usually scales the variables using characteristic quantities of the flow in order to obtain dimensionless variables. For wall bounded flows, as the ones considered here, the most characteristic quantity is the wall gradient γ_W (s.f. Sect. 5.9.2). With this quantity one defines the so called dimensionless wall distance y^+ as

$$y^+ = y \cdot \sqrt{\frac{\gamma_W}{\nu_S}}. \tag{5.77}$$

It is also common to introduce the so called wall shear velocity u_o via the relationship

$$u_o = \sqrt{\nu_S \gamma_W}. \tag{5.78}$$

Using this quantity leads to the following expression for the dimensionless wall distance:

$$y^+ = \frac{y \cdot u_o}{\nu_S}. \tag{5.79}$$

Velocities are made dimensionless by just dividing them by the wall shear velocity:

$$\boldsymbol{u}^+ = \frac{\boldsymbol{u}}{u_o}. \tag{5.80}$$

Times can also be scaled with the help of the wall gradient. So the dimensionless time, which can be used as the variable in the representation of temporal correlation functions is given by:

$$\tau^+ = \tau \cdot \gamma_W. \tag{5.81}$$

It can be shown, [15], that for well developed turbulent channel flows an analytic expression for the average velocity profile across the channel can be derived. This so called logarithmic wall law has the general form:

$$\overline{u}^+(y^+) = \frac{1}{\kappa} \ln(y^+) + C \tag{5.82}$$

where the constants have to be determined by comparison with experimental data. It turns out that $\kappa = 0,4$ and $C = 5$. For very small y^+, i.e. in the so called laminar wall layer very close to the wall, this law is not valid but one has instead

$$\overline{u}^+(y^+) = y^+. \tag{5.83}$$

In these equations we used the notation of Sect. 2.1 for the average and fluctuating part of the velocity.

There are two scalar quantities, which can be formed from products of velocity gradient components. One is the turbulent energy dissipation

$$\epsilon^+ = \sum_{i,j=1}^{3} \overline{\left(\frac{du_i^+}{dx_j^+} + \frac{du_j^+}{dx_i^+} \right)^2} \tag{5.84}$$

and the other the enstrophy

$$\omega^+ = \frac{1}{2} \sum_{i,j=1}^{3} \left(\frac{du_i^+}{dx_j^+} - \frac{du_j^+}{dx_i^+} \right)^2. \tag{5.85}$$

6. FRS: Forced Rayleigh Scattering

6.1 Basic Considerations

6.1.1 Introduction

Following a suggestion by de Gennes published in 1978 [16] Cloitre and Guyon [17] first measured velocity gradient values in a turbulent tube flow on the basis of the theory outlined here. The principle of their technique, which became known as 'forced Rayleigh scattering', will now be explained with the help of Fig. 6.1.

The beam of a pulse laser is split into two coherent beams, PL, of roughly the same intensity. These two beams are characterized by the wave vectors $\boldsymbol{k}_1$ and $\boldsymbol{k}_2$. They cross each other in the scattering volume V. This is the typical configuration of a laser doppler velocimetry set-up (c.f. Sect. 2.3.3). In the scattering volume, where the two illuminating light beams cross, one obtains an electric field by superposition during the time the pulse lasts. It is approximately given by

$$\boldsymbol{E}(\boldsymbol{r},t) = \boldsymbol{E}_o \cdot \mathrm{e}^{-i\omega t} \cdot \left(\mathrm{e}^{i\boldsymbol{k}_1\cdot\boldsymbol{r}} + \mathrm{e}^{i\boldsymbol{k}_2\cdot\boldsymbol{r}}\right) \tag{6.1}$$

The intensity in that region is then

$$\begin{aligned} I(\boldsymbol{r},t) &= \boldsymbol{E}(\boldsymbol{r},t)\cdot\boldsymbol{E}^*(\boldsymbol{r},t) \\ &= |\boldsymbol{E}_o|^2\cdot\left(\mathrm{e}^{i\boldsymbol{k}_1\cdot\boldsymbol{r}} + \mathrm{e}^{i\boldsymbol{k}_2\cdot\boldsymbol{r}}\right)\left(\mathrm{e}^{-i\boldsymbol{k}_1\cdot\boldsymbol{r}} + \mathrm{e}^{-i\boldsymbol{k}_2\cdot\boldsymbol{r}}\right) \\ &= |\boldsymbol{E}_o|^2\cdot\left(2 + \mathrm{e}^{i\boldsymbol{q}\boldsymbol{r}} + \mathrm{e}^{-i\boldsymbol{q}\boldsymbol{r}}\right) \\ &= 2\cdot|\boldsymbol{E}_o|^2\cdot(1 + \cos(\boldsymbol{q}\cdot\boldsymbol{r})) \end{aligned} \tag{6.2}$$

where we introduced the difference vector

$$\boldsymbol{q} = \boldsymbol{k}_1 - \boldsymbol{k}_2. \tag{6.3}$$

It can be seen in the last part of the equation (6.2), that the intensity has the form of a plane time independent 'intensity wave'. Because it is a real wave, it is composed of two waves with wave vectors $\boldsymbol{q}$ and $-\boldsymbol{q}$, which only differ in the sign. In the figure, only one of these wave vectors is shown.

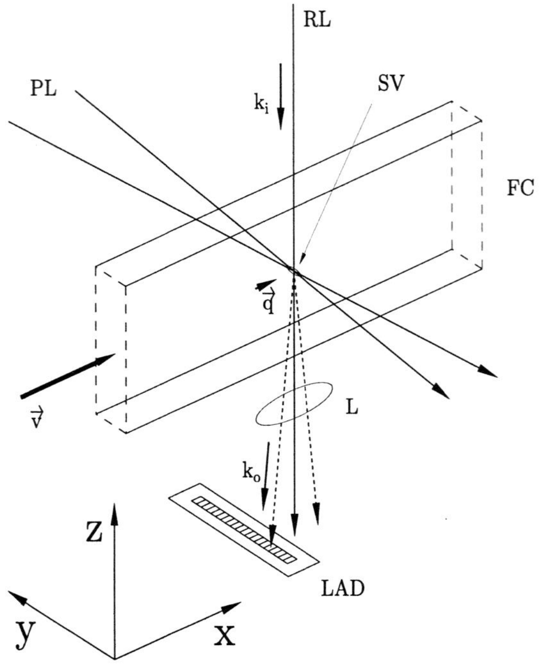

Fig. 6.1. Schematic of the experimental set-up described in [17] in the vicinity of the scattering volume. Two beams of a pulse laser, PL, and the beam of the probe laser, RL, cross each other in the scattering volume, SV, within the flow tube, FC. The diffracted light of the probe laser is focused by the lens L, onto the linear array detector, LAD.

If a certain amount of the light energy is absorbed within the fluid it will be transformed to heat, and therefore a stationary 'temperature wave' is produced within the fluid. Generally, fluids expand on raising the temperature. Consequently, the temperature wave will entail a density wave. The time necessary to produce this density wave is extremely short: it is roughly equal to the wavelength of this wave divided by the velocity of sound. The refractive index is dependent upon the density of the medium, and we finally end up with a refractive index wave within the fluid, which can also be considered as an optical three dimensional grating. The light beam of a probe laser of low intensity can be diffracted by this grating. One of the diffracted light waves with the wave vector $\boldsymbol{k}_a = \boldsymbol{k}_e - \boldsymbol{q}$ will then be detected by the linear array detector LAD. Due to the high thermal conductivity of the fluid, the grating will not last very long and, thus, the diffracted light wave of the probe laser will disappear quickly.

In the case of a non-vanishing velocity gradient within the region where the mentioned grating is produced the diffracted light wave will generally change in time. This change is dependent on the change of the wave vector

$\boldsymbol{q}$ which is described by equation (4.65). It is obvious that the two wave vectors $\boldsymbol{q}$ and $-\boldsymbol{q}$ stay always antiparallel. Consequently, as long as the optical grating exists it will remain a pure sinusoidal grating, with only its orientation in space and its wavelength subject to changes. Thus, the direction of the diffracted light will change, which can be measured by use of the linear array detector. It is then possible to calculate one particular component of the velocity gradient tensor from the measured velocity of the observed diffraction spot.

6.1.2 The Lifetime of the Grating

An experimental technique based on the principle described in the last section hinges on a sufficiently long life time of the refractive index grating. It must be long enough so that the change in the direction of the diffracted wave can easily be measured. This is especially a problem with small gradients.

The life time of the grating, when no velocity gradient is present, is given by

$$\tau_G = \frac{D_T}{\boldsymbol{q}^2}, \tag{6.4}$$

where D_T is the thermal diffusion coefficient.

The velocity within the scattering volume influences the accuracy of the method because it lessens the life time of the grating within the spatially fixed scattering volume.

The life time of the grating can be increased considerably by dissolving a bleachable dye within the fluid. The dye molecules are excited by the light of the pulse laser to a metastable electronic state. Molecules in that state can no longer absorb light of the probe laser. One then has an absorption grating, which not only has a much longer life, but which also has a much higher diffraction efficiency. The longer lifetime furthermore allows a reduction in the lattice constant of the diffraction grating by increasing the angle of the two coherent light waves of the pulse laser. This entails a larger diffraction angle of the probe laser light and with it a higher accuracy. Gradient measurements using a bleachable dye are described in detail in [18].

The lattice constants of the optical gratings which were used for gradient measurements were in the range of 10 -70μm. The angle of the two light waves of the pulse laser is then in the range of a few degrees. The time period which is used to follow the motion of the diffracted probe laser light was in the range of $\approx$100μs to a few milliseconds.

6.1.3 Scattering Geometry and Scattering Volume

The scattering volume, by this measurement technique, is determined by the crossing of three light beams: The two pulsed beams in laser-doppler geometry and the probe beam. The two pulsed beams determine the vector

$\boldsymbol{q}$ on the right hand side of equation (6.2). Consequently they determine that the derivatives of the component in direction of $\boldsymbol{q}$ of the flow velocity can be measured. Because of the rather large lattice constant of the diffraction grating the probe laser beam must be almost perpendicular to the vectors $\pm\boldsymbol{q}$. One then has two main directions for the probe beam:

- probe beam perpendicular to the plane defined by the two pulsed beams
- probe beam in the plane of the two pulsed beams

The first scattering geometry has the advantage that the scattering volume can be made fairly small. Shaping of the laser beams to form elliptical cross sections leads to the required scattering volumes demanded by the flow. Cloitre and Guyon [17] have thus formed an oblate scattering volume which has its short axis in the direction of the probe laser beam. The resulting prolate q-volume of coherence then suppresses delusive speckle motion. (For a detailed explanation see Sect. 4.7.3).

The possibilities of shaping the scattering volume are restricted for the second scattering geometry. Due to what was said above, the angle between the probe beam and the two pulsed beams must be very small. This leads to a rather extended scattering volume in the direction of the probe beam, which cannot be reduced significantly.

It is possible to measure all three spatial derivatives of a certain velocity component if two probe beams in different directions are used and the motions of the respective diffracted light beams are measured. All gradient components require measurements with three different directions of the pulsed beams.

Due to the rather large lattice constant of the optical lattice the angle of diffraction of the probe beam is fairly small, which of course restricts the sensitivity of the technique.

6.2 Experiments and Results

So far two publications have appeared which show results of measurements of gradients in fluid flows obtained with the technique described in this chapter: [17], [18]. We will discuss here the work by Cloitre and Guyon [17] which contains many more experimental results than the other publication.

6.2.1 Experimental Set-Up

The principal part of the experimental set-up is shown schematically in Fig. 6.1. The two pulsed laser beams are produced by a Nd:YAG-laser with a pulse power of 10 mJ and pulse duration of 10 ns. Both beams are given an elliptical profile by use of cylindrical lenses so that the long axis of the beam profiles lie in the plane defined by the two beams. The probe beam supplied by a HeNe-laser crosses the two other laser beams in their overlap region normal to the just mentioned plane. The lattice constant of the optical lattice can be varied by changing the angle between the two pulsed beams. The linear CCD-array is placed in the focal plane of the lens L. The size of the scattering volume is $300\times300\ \times50\mu m^3$. The flow tube has a rectangular cross section of the size 5mm×40mm, the scattering volume was placed 700 mm away from the inlet.

The motion of the diffracted light beam in this scattering geometry is determined by the two gradient components Γ_{xx} and Γ_{xy}. The first component determines the motion in the x-direction, the second one that in the y-direction. Assuming that the first component is small compared to the second - which is obviously the case near the wall - the main motion of the diffracted light beam is in the y-direction. For this reason the authors have oriented the linear array detector parallel to the y-axis. If the light intensity is measured with the detector at a definite time after the light pulse of the Nd:YAG-laser, one obtains a series of intensity data with a clearly marked maximum. From its position, the time lag, the focal length of the lens L, and the distance of the pixels one can easily calculate the rate of change of the diffracted light, i.e. the y-component of $\dot{\boldsymbol{k}}_o$.

6.2.2 Experimental Results

For well developed turbulence the gradient profile of Fig. 6.2 is obtained.

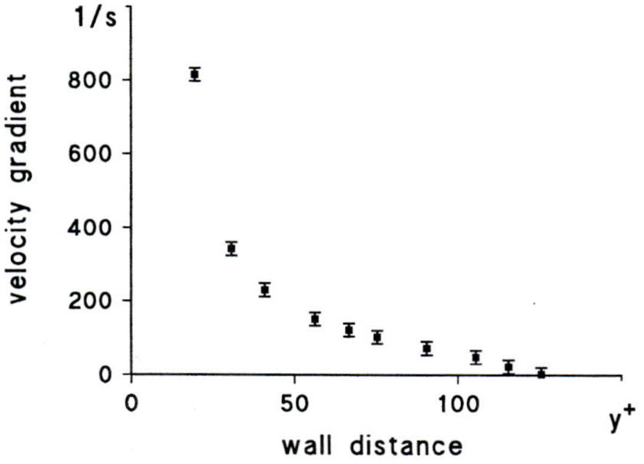

Fig. 6.2. Profile of the gradient component Γ_{xy}. The dimensionless wall distance y^+ is calculated according to equation (5.79).

Figure 6.3 shows the average gradient for three different Reynolds numbers below and above the critical Reynolds number.

Since this technique yields instantaneous gradient values, one can measure gradient distributions by repeated measurements. Figure 6.4 shows results of measured gradient distributions for different wall distances.

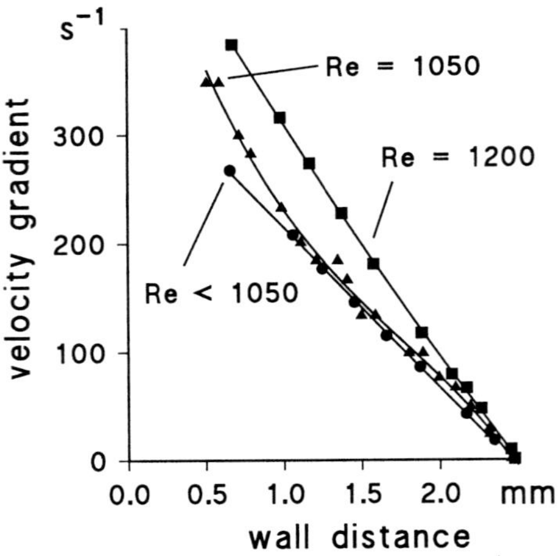

Fig. 6.3. Dependence of the average gradient Γ_{xy} on the wall distance. The measurements give a critical Reynolds number of ≈ 1050.

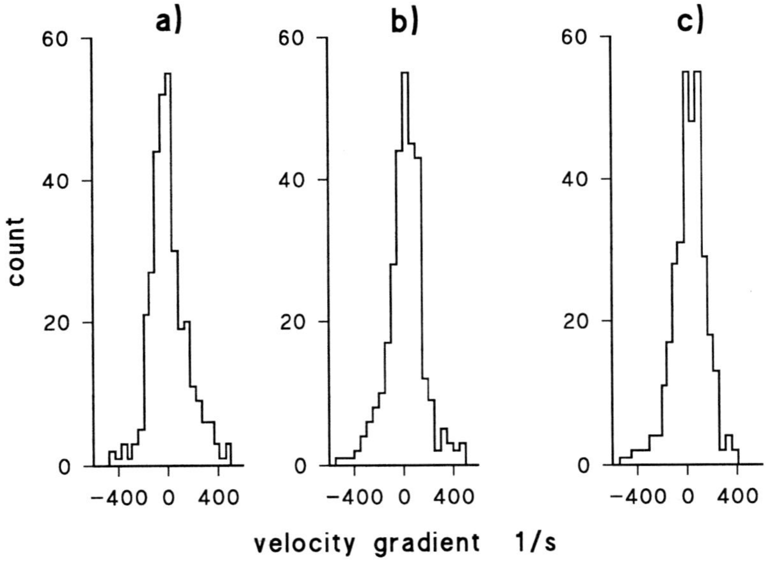

Fig. 6.4. Gradient distributions obtained on the basis of 300 single measurements for three different wall distances: a: $y^+ = 80$, b: $y^+ = 120$ (centre of flow tube), and c: $y^+ = 160$.

7. LICC: Light Intensity Cross-Correlation

7.1 Basic Considerations

7.1.1 Introduction

This measurement method, as well as the others described in the following sections, demand a seeding of the flow under investigation (c.f. Chap. 2.3.1). The result is a stochastic pattern of the scattered light (see Sect. 5.3). The basic idea of the method described in this chapter can be understood very easily if a special case is considered: let us assume that the pattern of the scattered light moves rigidly and the motion direction is known. Under these condition the pattern velocity can be measured in the following way: Two light detectors are placed in the plane of observation, such that they lie on a line parallel to the pattern motion direction. One then forms the temporal correlation function of the detector signals. If the distance of the two detectors is called d and the velocity of the pattern v_P, then the light intensity at the first detector at a certain time t is evidently the same as that at the second detector at time $t + d/v_P$. This gives the following relationship between the intensities at the two detectors:

$$I_2(t + d/v_P) = I_1(t). \tag{7.1}$$

From this equation follows immediately

$$I_2(t) = I_1(t - d/v_P). \tag{7.2}$$

It is then possible to express the cross-correlation function

$$C_{12}(d, \tau) = \langle I_1(0) I_2(\tau) \rangle \tag{7.3}$$

by the auto-correlation function. For that purpose we replace $I_2(\tau)$ in equation (7.3) with the help of equation (7.2) and obtain

$$C_{12}(d, \tau) = \langle I_1(0) I_1(\tau - d/v) \rangle = C_{11}(0, \tau - d/v_P). \tag{7.4}$$

The auto-correlation function on the right hand side has an absolute maximum if all its arguments are zero (see equation 3.47), i.e. for $\tau - v_P/d = 0$

which amounts to $\tau = d/v_p$. Consequently, the cross-correlation function also has an absolute maximum for this value of τ. If, therefore, the temporal position of the maximum of the correlation function, τ_{max}, is determined, it can be used, together with the distance of the detectors, d, to calculate the velocity of the pattern. However, this only leads to a correct result if the pattern moves rigidly, i.e. it exhibits no boiling. Obviously, this method may also yield acceptable results, if there is a slow boiling of the pattern, and if the motion direction of the pattern is not exactly parallel to the line connecting the detectors, but it may be hard to prove the reliability of the obtained data.

7.1.2 The Demands on the Experimental Set-Up

As became clear from the discussion of the last section, one has to ensure that the most important requirements for a successful application of this method are met:

- both detectors must lie on a speckle path
- the pattern must have a life time which exceeds the time necessary for it to travel the distance between the two detectors.

These requirements seem to restrict a possible application of the method to laminar flows. However, we are going to show that this method can be applied to a turbulent channel flow. Let us consider a laminar flow where the average value of one off-diagonal element of $\hat{\Gamma}$ only is not equal to zero. This is for instance the case in the horizontal mid-plane in a rectangular flow tube (as shown in Fig. 7.1) or in an open rectangular channel far away from the bottom. In the co-ordinate system shown in the figure this component is $\frac{\partial v_x}{\partial y}$. Immediately at the wall this gradient - the wall gradient - has its maximum value, and it drops down to zero on going to the middle of the flow.

If the flow becomes turbulent on increasing the flow velocity the other components of the gradient tensor will no longer be zero. They will, however, have an average value of zero. So let us assume that up to a certain distance from the wall the speckle motion is mainly determined by the velocity gradient $\frac{\partial v_x}{\partial y}$. The situation is therefore almost the same as the one discussed in Sect. 5.9.2. The scattering geometry which is suited best for the measurement of $\Gamma_{xy} \neq 0$ is shown in part a) of figure 7.1: The scattering vector $\boldsymbol{q}$ points in the direction of the average flow velocity and the speckle motion caused by this gradient to the y-axis. Due to the fact that the direction of the speckle motion is perpendicular to the scattering vector $\boldsymbol{k}_o$, no oblique speckle motion occurs.

If now the other velocity gradient components are not zero, but small, boiling and oblique speckle motion will occur, but their influence will be small.

Unfortunately, the measurements we performed were done with an open water channel where the above scattering geometry is not possible. We are

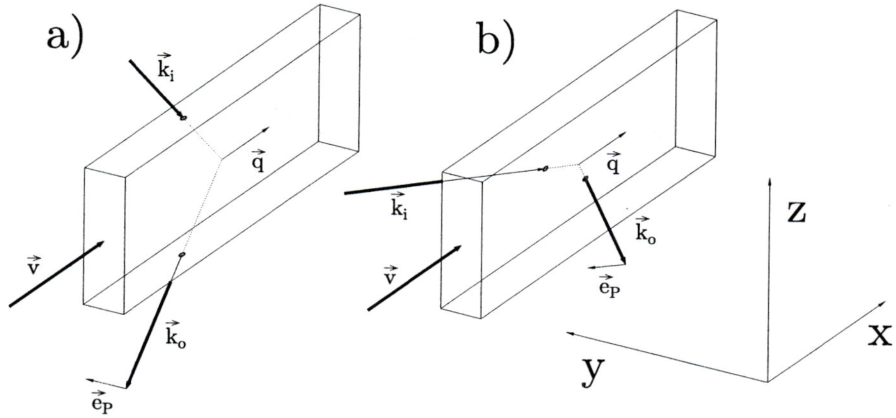

Fig. 7.1. a): The optimum scattering geometry for the measurement of Γ_{xy}, b): The scattering geometry to measure $\Gamma_{xx} - \Gamma_{xy}$.

therefore going to consider a scattering geometry which can be used in this case. This is shown in part b) of the figure. Here too, the scattering vector is along the x-direction. However, the vector $\boldsymbol{k}_o$ is not perpendicular to the y-direction, which means that the speckle motion in the y-direction cannot be measured directly. Consequently, we have to deal with oblique speckle motion in this case (see Sect. 4.7.1). With an angle of 45° between the vectors $\boldsymbol{k}_o$ and $\dot{\boldsymbol{k}}_o$, the boiling of the pattern will be fairly strong. One way to considerably reduce the boiling and obtain nevertheless rather long speckle paths, is to shape an oblate scattering volume as described in Sect. 8.2. The resulting prolate q-volume of coherence causes the desired effects.

Using the scattering geometry shown in Fig. 7.1 b), one does not measure the component Γ_{xy}, but the linear combination $\Gamma_{xx} - \Gamma_{xy}$. This follows from a direct comparison with the situation discussed in Sect. 5.8.2. In case of the channel flow, however, one has

$$\left\langle \frac{du}{dx} \right\rangle = 0 \tag{7.5}$$

and therefore

$$\left\langle \frac{du}{dx} - \frac{du}{dy} \right\rangle = - \left\langle \frac{du}{dy} \right\rangle . \tag{7.6}$$

It is thus possible to measure the mean wall gradient directly with this set-up, i.e. the quantity $\left\langle \frac{du}{dy} \right\rangle$.

Section 7.2.1 describes another way to apply this method in the case of oblique speckle motion. One can generally state that this method can be used to measure average gradients whenever two conditions are - at least approximately - fulfilled:

- One gradient component is much bigger than all others. Then the vector $\dot{\boldsymbol{k}}_o$, which governs the speckle motion, has a rather well defined direction.
- It is possible, by choice of the scattering geometry, to shape the scattering volume, or by some other means, to sufficiently reduce the speckle boiling.

7.1.3 Evaluation Using Fit Procedures

For increasing distances of the detectors, i.e. for larger d, or for speckle motions which do not exactly coincide with the line connecting the two detectors, one will observe that the maximum of $C_{12}(d, \tau)$ will become smaller and smaller in comparison to the maximum of $C_{11}(0, 0)$. In the first case, the pattern boils away on its path from the first to the second detector, in the second case, any speckle which hits the first detector, partly misses the second. Both effects cause the position of the maximum of the cross-correlation function to move to a smaller τ. This means that the pattern velocity will be overestimated if the position of the maximum is used for the calculation of the pattern velocity. If the direction of the pattern velocity is not very well known, the movement of the maximum of the cross-correlation function will generally influence the measured data and, therefore, this simple type of data evaluation cannot be recommended.

As the results shown below indicate, one can successfully fit the measured correlation function to a rather general function with free parameters. The form of this function will be based on a theoretical modeling of the light scattering theory, as well as on general assumptions on the distribution of the velocity gradient within the flow. The advantage of being free of a model in the evaluation of the data is of course sacrificed, which definitely limits the applications of this method.

The above procedure was used by the author ([19]) whose experiments we are now going to consider. We will not present here the rather tedious calculation of the correlation function, which can be found in [11]. The chief assumptions for this calculation are the following:

- The light intensity distribution within the scattering volume is Gaussian.
- The diaphragms in front of the detectors are circular.
- The distribution function of the component of the velocity gradient in the turbulent flow under consideration is also a Gaussian.

With these assumptions and a few approximations one can derive the general form of the correlation function:

$$C_{12}(\tau) = A_1 + A_2 \mathrm{exp}(A_3\tau^2)\mathrm{exp}(A_4\tau)\mathrm{exp}(A_5|\tau|)\mathrm{exp}(A_6\tau^3)\mathrm{exp}(A_7\tau^4). \quad (7.7)$$

The important parameter for the determination of the velocity gradient is A_4. The two parameters A_6 and A_7 result from the fluctuating gradients in the turbulent flow, both are zero for laminar flows. A_5 depends on the diffusion of the light scattering particles, and A_3 is determined by the particle velocity and those gradient components, which cause boiling of the pattern.

7.2 Experiments and Results

The method described here was developed by two students working on their thesis, [20] and [19]. We will discuss here the experimental set-up developed by the second author and show some of his results.

7.2.1 Experimental Set-Up

The intention of the author was to measure the time behaviour of the gradient component Γ_{xy} in a flow channel near the wall, i.e. he had to measure the speckle motion temporally resolved. He had to take into account that the two gradient components Γ_{xx} and Γ_{xz} also determine the speckle motion; they only have a vanishing temporal average value. For that reason he has modified the set-up shown schematically in Fig. 7.1 b) as follows:

Instead of an illuminating light beam with a fixed direction, he used a beam that could be switched between two different directions. If these directions are chosen properly one can measure directly the desired gradient component Γ_{xy}. This should be explained with the help of Fig. 4.7:

Two light beams with wave vectors $\boldsymbol{k}_{i1}$ and $\boldsymbol{k}_{i2}$, illuminate alternatively the scattering volume. The directions of these vectors, as well as the vectors $\boldsymbol{k}_{o1}$ and $\boldsymbol{k}_{o2}$, are chosen in such a way, that $\boldsymbol{k}_{i1}$ points in the x-direction and $\boldsymbol{\kappa}$ in the y-direction. The angle between the two illuminating light beams is heavily exaggerated in the figure, it should be only a few degrees. If one aligns the optical elements carefully, the spatial aperture functions can be assumed to be identical. Now the detectors are gated in such a way that the electronics receive signals of detector 1 (the one in direction of $\boldsymbol{k}_{o1}$) only if the illuminating laser light shines in direction 1 (characterized by the wave vector $\boldsymbol{k}_{i1}$), and vice versa. The cross-correlation function of the signals of the two detectors is then - apart from a constant factor - equal to $C_I(\boldsymbol{q}_1, \boldsymbol{q}_2, \tau)$ in the notation of Sect. 4.6. The measured component of the speckle velocity is in the direction of $\boldsymbol{\kappa} = \boldsymbol{q}_1 - \boldsymbol{q}_2$, which is identical to the y-direction. Consequently, the velocity gradient component that is measured is exactly the required Γ_{xy}. The switching between the two directions of the illuminating laser beam must of course be fast enough. Only if the speckle velocity can be considered constant during the time period needed to switch the beam back and forth, can the obtained speckle velocity be considered reliable.

The experiments were performed with a 25 mW HeNe-laser as the light source and an open water channel with dimensions 500cm×18cm×3cm. Photomultipliers served as detectors, and the correlation functions were calculated by a 4-bit digital correlator. The electronics of the correlator were slightly modified to allow the above mentioned gating of the detector signals. The switching of the laser beam directions was achieved by use of a Pockels cell, and the digital correlator and Pockels cell were synchronized so that exactly one Pockels cell cycle ended during one sample time of the correlator.

The parameters of the function given in equation (7.7) were determined by least mean square fits to the measured correlation functions. The set-up shown in Fig. 7.1 b) gives

$$\left\langle \frac{du}{dy} \right\rangle = -\frac{A_4}{\Phi_{11} \cdot Q \cdot \Delta q} \tag{7.8}$$

where

$$Q = \frac{1}{2}|\boldsymbol{q}_1 + \boldsymbol{q}_2| \tag{7.9}$$

$$\Delta q = |\boldsymbol{q}_1 - \boldsymbol{q}_2| . \tag{7.10}$$

The 3×3 tensor $\hat{\Phi}$ is a function of the size of the scattering volume and the diameter of the circular detector diaphragms. It is given explicitly in [11].

For the small detector diaphragms (diameter smaller than 0.3 times the diameters of the coherence area) used here, and with the assumption of a spatial aperture function of the form

$$(\boldsymbol{r}) = A \cdot \exp(-x^2 \Sigma_x - y^2 \Sigma_y - z^2 \Sigma_z), \tag{7.11}$$

one obtains for the desired component of $\hat{\Phi}$

$$\Phi_{11} = \frac{1}{2}\left(\Sigma_x + \Sigma_y\right) . \tag{7.12}$$

7.2.2 Experimental Results

Figure 7.2 shows a measured correlation function together with the determined fit function of the type (7.7).

Figure 7.3 shows two examples of measured gradient profiles. Deviations from the expected behaviour near the wall, where the gradient is supposed to be constant, can clearly be seen. This behaviour of the experimentally determined values is due to the fact that at small wall distances the scattering volume is cut by the glass wall of the flow channel. Scattering particles, which are adsorbed at the wall, provide a constant background of coherent stray light (see Sect. 5.5), which reduces the value of the determined pattern velocity. Furthermore, the scattering volume will no longer be Gaussian, which renders the evaluation procedure dubious.

The explanation just given is proven by measurements with altered parameters in the optical set-up. Figure 7.4 shows results of measurements with different diameters of the scattering volume. It is obvious that the data obtained with the smaller scattering volume is reliable to much smaller wall distances than that with the larger scattering volume. Further experimental results obtained with this method are shown in the figure 9.6.

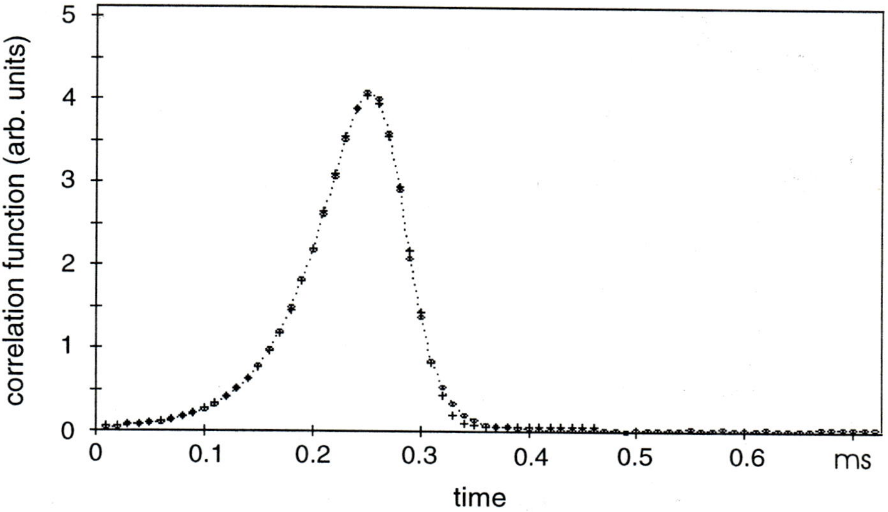

Fig. 7.2. Measured cross-correlation function at a Reynolds number of 3000. • : measured data, + : fitted data according to (7.9).

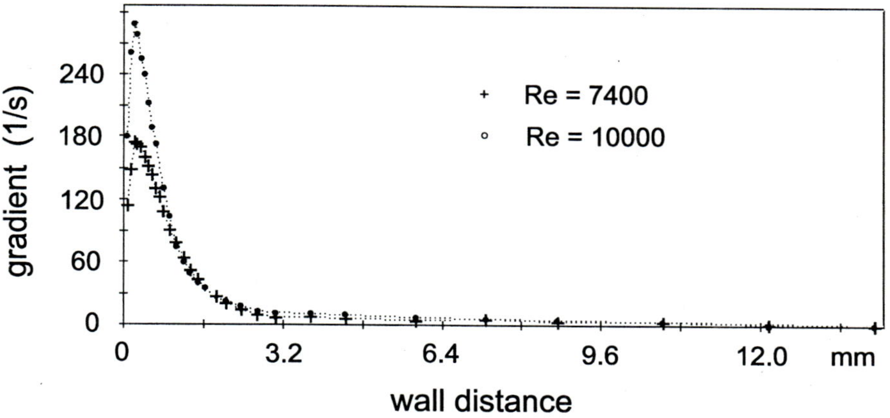

Fig. 7.3. Measured gradient profiles of $\Gamma_{xy}(y)$. + : Re = 7400; • : Re = 10000.

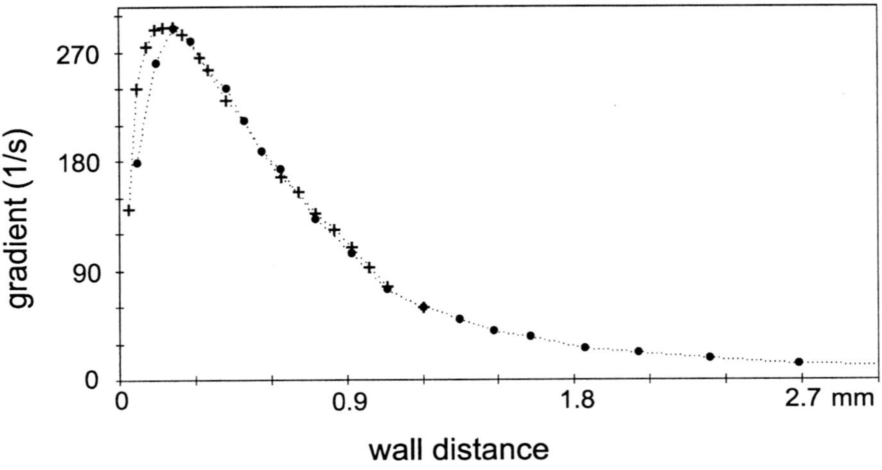

Fig. 7.4. Dependence of the experimentally determined velocity gradient in the near wall region on the diameter of the scattering volume at Re = 10000. Diameter of the scattering volume : + : 75μm, • : 150 μm.

8. DPSS: Double Pulse Speckle Strophometry

8.1 Basic Considerations

8.1.1 Introduction

This measurement technique applies spatial averaging for the determination of the pattern velocity. For that purpose, the flow is illuminated by two successive laser pulses of short duration, with a short temporal distance Δt_c. Both resulting speckle patterns are recorded electronically. Provided the speckle boiling is not too pronounced during the time Δt_c, the two pictures can be used to determine the motion of the pattern. One method to obtain the shift $\Delta \boldsymbol{\rho}_c$ of the pattern during the time Δt_c is to calculate the spatial cross-correlation function of the two images and determine the position of its maximum. The time Δt must of course be chosen so short that the maximum in the cross-correlation function is well developed and can be discriminated against the random maxima occurring above the background, which are caused by the finite area of the images. The pattern velocity is given by the two quantities just introduced via

$$\boldsymbol{v}_P = \frac{\Delta \boldsymbol{\rho}_c}{\Delta t_c}. \tag{8.1}$$

The implementation of this technique requires the consideration of a number of different aspects which will now be discussed in detail. As far as the detector is concerned we will restrict our consideration to a linear array detector. A detector of that type was used for our experiments.

8.1.2 The Admissible Size of the Recorded Speckle Pattern

According to equation (4.71), the velocity of the speckle pattern varies with the direction of the scattered light, i.e. with $\boldsymbol{k}_o$. This means that the speckle velocity is not constant over the area of the images discussed in the last section. For that reason, one can only speak in an approximate sense of the pattern velocity over an area on the observation screen. How good this approximation is, or how big the area is over which the velocity is fairly constant, cannot be decided generally because it depends on the spatial form of the velocity gradient tensor.

It is possible to estimate the influence of this effect on the obtained results under some simplifying assumptions. Let us assume that all elements of the gradient tensor are of the same order of magnitude. Let us furthermore restrict our consideration to one dimension only. If the detector has an extension so that it covers an angle of 0.1 rad of the scattered radiation, then the variation in pattern velocity along this direction is about 3%. However, this does not entail an error of the same order of magnitude, because the pattern motion is calculated from the intensity values across the whole image area. So the effect on the cross-correlation function will generally be a broadening of the maximum so that the shift of the maximum will be very much smaller than the 3%. Therefore, an error caused by this effect can be ignored under these circumstances. The measurements shown in this chapter are made with a detector of the spatial extension just considered.

8.1.3 The Necessary Speckle Number in the Recorded Speckle Pattern

It is obvious that for the calculation of a reliable cross-correlation function, the averaging must be made over a sufficient number of coherence areas, or, in other words, the number of speckles within in the recorded image must be large enough. Only if this is the case can one discriminate the 'true' maximum in the correlation function against 'random' ones.

In the experiments described below, which were performed with a linear array detector, an averaging over about 50 diameters of the coherence area turned out to give satisfactory results. If the speckle number should turn out to be too small one can increase it by a number of means. The scattering volume may be increased by use of a lens of smaller focal length in the illuminating light path, which reduces the size of the coherence area. It is also possible to reduce the size of the speckles on the detector by altering the optics on the detection side.

8.1.4 Oblique Speckle Motion and the Shape of the Scattering Volume

For an accurate determination of the position of the maximum in the correlation function, which is required to obtain reliable gradient values, one has to make sure the path the pattern traveled in the observation plane is large enough compared with the diameter of the coherence area. If, however, the speckle path $\Delta\rho_c$ is considerably larger than the speckle size one will have difficulties because of oblique speckle motion: In that case the patterns of the two images will show no correlation whatsoever because the pattern recorded first will move out of the Ewald sphere.

A possibility to deal with this difficulty is to use an optical set-up which causes a prolate q-volume of coherence where the large axis is parallel to the vector $\boldsymbol{k}_o$. Then one will always observe the pattern velocity components, which are perpendicular to $\boldsymbol{k}_o$, regardless of the speckle motion being oblique or not. This can easily be seen with the help of Fig. 8.1 (see also Fig. 4.8).

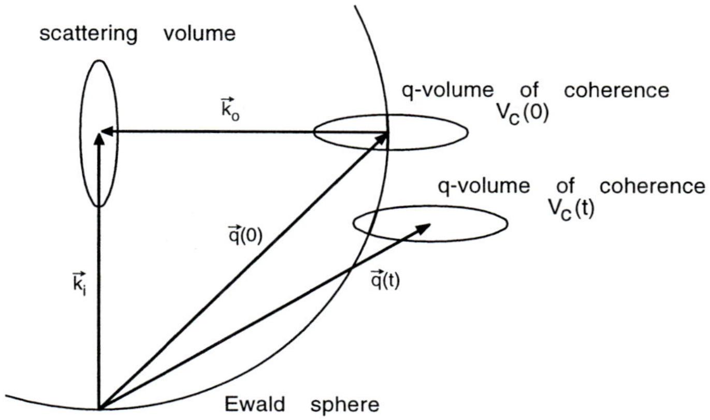

Fig. 8.1. Speckle motion in k-space and resulting motion of scattered light pattern with elliptical scattering volume

With the resulting prolate q-volume of coherence, the pattern stays coherent fairly long, even with oblique motion. The necessary oblate scattering volume can be shaped by use of cylindrical lenses (see Sect. 8.2.1 and figure 8.2).

8.1.5 Special Demands on the Optical Set-Up

This technique requires laser pulses which are considerably shorter than the time span between the two pulses in order to obtain images which are not degraded. It is therefore advisable to use a double pulse laser. The required light energy of a pulse is of the order of 10 mJ. The pulse duration is usually on the order of 10 ns, which is smaller than actually required. Due to the high energy density, which results from these parameters, problems may arise especially in those regions where the diameter of the laser beam is very small. That happens at the pin hole of a spatial filter and near the scattering volume. Measurements near the wall of a flow vessel can be critical since small particles adhering to the wall may absorb a fraction of the light large enough to lead to damages of the wall.

The light sensitive areas of a linear array detector - the so called pixels - are very small (see the following section). We used a detector with a pixel size of 13x13 μm. This requires a very reliable adjustment to focus the speckle pattern, which is made quasi one-dimensional on the linear array detector by use of a cylindrical lens. High precision mechanical elements are required for this, so that linear adjustments can be made reproducibly to within a few microns, and optical elements can be rotated to a comparable accuracy.

8.1.6 The Electronic Recording of the Speckle Pattern

At the time when the experiments cited here were performed, no electronic devices were yet available which allowed for the recording of two two-dimensional images to within a few microseconds. We therefore decided to 'compress' the speckle pattern in one direction with a cylindrical lens to obtain a quasi one-dimensional pattern. Linear array detectors, which meet the just mentioned demands, are available. They furthermore reduce considerably the effort to calculate the pattern velocity, although they yield only one velocity component of the pattern.

A single diode usually has a quadratic cross section; it delivers an electric charge proportional to the light energy absorbed by it. The charges built up in the single diodes are then successively shifted with external clock pulses and amplified by external electronics. One finally has a digitized image of the light intensity. Obviously, a reliable image of the intensity pattern can only be obtained if the pixel size is small compared to the diameter of the coherence area in the direction of the array detector. Our measurements show that the number of pixels within a coherence length should not be smaller than 4, otherwise, not only the accuracy decreases, but there will also occur *systematic* deviations in the determined pattern velocities.

Two properties of real CCD-array detectors may influence the obtained results and will therefore be discussed. Firstly, the sensitivities of the single diodes are not exactly equal, and sometimes one observes also a fairly smooth variation of the intensity along the detector. Usually, the differences between neighbouring diodes are small and need not be considered. A variation across the detector may occur for a variety of reasons. It is possible that the angle, which the illuminating light forms with single diode planes, varies because of an irregular plastic cover of the array detector. Scratches, dust or fingerprints can also lead to different diode signals.

Secondly, the array detector used had the special property that, after the first illumination, the charges of the single diodes could be - triggered by an external pulse - transformed to storage capacitors. This transfer of the charges is not perfect, i.e. a small amount of charge is left. That means that after the second illumination the charge of a diode is not independent of the intensity of the first illumination. Therefore the second image contains a certain fraction of the first image. We found that when this fraction was smaller than 5% one can neglect this effect.

8.1.7 The Determination of the Maximum of the Cross-Correlation Function

In this technique spatial averaging over a fairly small fraction of the pattern is performed. Consequently, it is far from ideal. For this reason, one observes in any calculated cross-correlation function of two images a number of maxima and minima of different height. One cannot simply state that the largest maximum is caused by the shift of the pattern, whereas the other ones are random. It is therefore necessary to find a criterion which discriminates between the 'true' and the 'random' maxima. For that purpose we introduced the following normalization of the cross-correlation function:

$$C(j,\tau) = \frac{\frac{1}{N-j}\sum_{i=1}^{N-j} S_1(i+j,0)S_2(i,\tau)}{\sqrt{\langle S_1^2(j)\rangle \langle S_2^2(j)\rangle}} \tag{8.2}$$

where

$$\langle S_1^2(j)\rangle = \frac{1}{N-j}\sum_{i=j+1}^{N} S_1^2(i,0) \tag{8.3}$$

and

$$\langle S_2^2(j)\rangle = \frac{1}{N-j}\sum_{i=1}^{N-j} S_2^2(i,\tau). \tag{8.4}$$

Here $S_1(i,0)$ and $S_2(i,\tau)$ denote the signals of the i-th pixel at time 0 and τ, respectively. We call the maximum of this normalized correlation function 'correlation coefficient'. A systematic investigation showed that this coefficient should be larger than 0.5 to be sure that the corresponding maximum is the one caused by pattern motion. While the measurements are performed, one has to make sure that only on rare occasions the correlation coefficient drops below the critical value 0.5. Otherwise there is a risk that especially large gradient values are suppressed and quantities calculated from the measured gradients will be biased.

If the 'true' maximum has been found by this method one has to determine its position. This can be done with much higher precision than might be suspected on the basis of the pixel distance. We fitted to the upper part of the maximum a parabola and determined the maximum position of this parabola. By this method we could determine the position of the maximum to an accuracy of about 1/7 of a pixel distance.

8.2 Experiments and Results

The method described was developed during the work required for a thesis [21]. All results shown here are taken from this thesis.

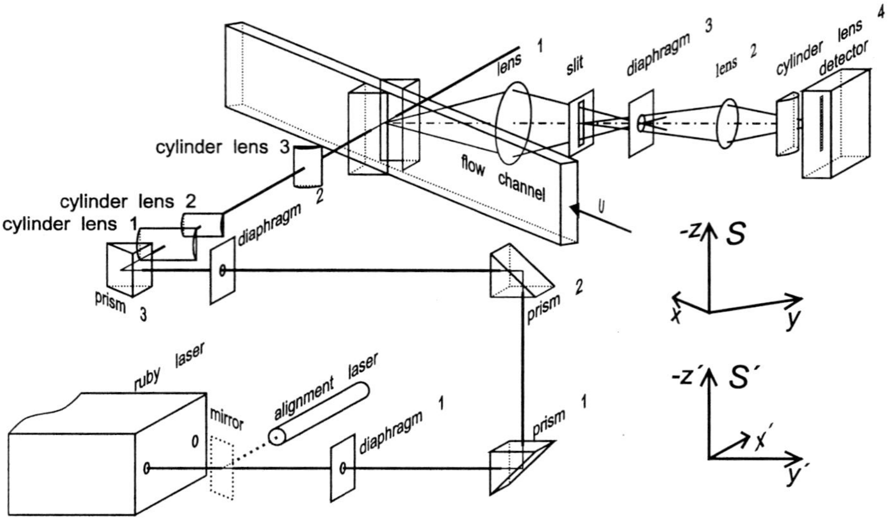

Fig. 8.2. Optical set-up of the double pulse method. Explanation in the text.

8.2.1 Experimental Set-Up

The optical set-up is shown schematically in Fig. 8.2. The cylindrical lenses 1, 2, and 3 serve to produce an elliptical cross section of the illuminating laser beam. The vertical diameter of the beam in the scattering volume was 1mm and its horizontal diameter 35μm. The function of the cuvettes attached to the outside of the flow tube is explained in Sect. 5.8.1. The guidance of the scattered light differs form that shown in Fig. 5.3 by a number of added components. The additional cylindrical lens serves to focus the scattered light in the horizontal direction onto the CCD-array. The light is not affected in the vertical direction. The image of a cylindrical lens is of rather poor quality for rays, which form larger angles with the optical axis in the plane where the curvature of the lens is maximal. We therefore restricted the light reaching this lens by a rectangular diaphragm, which in the figure is called 'slit'. The imaging of the bright laser beam within the flow by lens 1 onto the plane of the diaphragm 3, restricts the scattering volume laterally to 400μm. The resulting effective scattering volume had the following dimensions with respect to the axes x',y', and z' of 400μm, 35μm and 400μm, respectively.

The rectangular flow tube has a cross section of 10.2mm$\times$80mm and a length of 150cm. The location of the scattering volume was 120 cm downstream from the inlet. Further information is found in [22].

8.2.2 Experimental Results

Figure 8.3 shows examples of two one-dimensional intensity images as were recorded with the linear CCD-array detector. The temporal distance of the

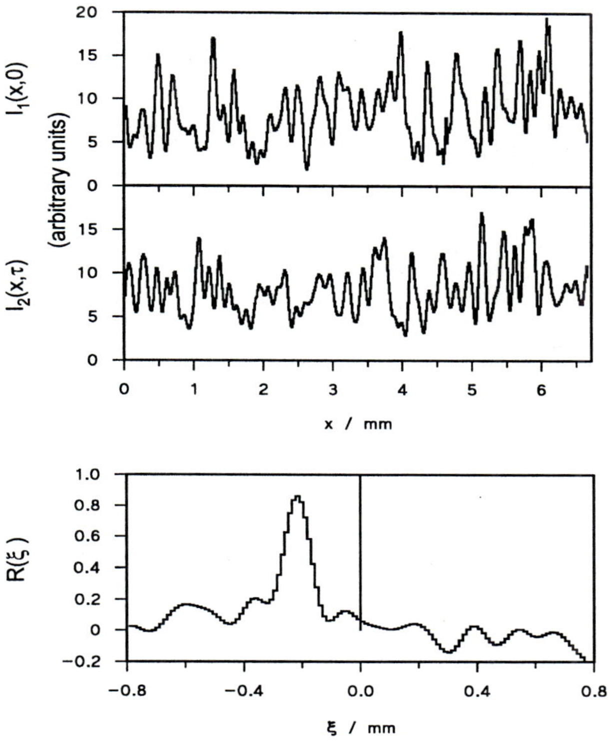

Fig. 8.3. Two intensity series obtained with a linear CCD-array detector. Time lag between the images 50μs. The lower part shows the calculated correlation function with a changed time scale.

two pictures was 50μs. In the lower part of the figure, the central section of the cross-correlation function of the two upper images is shown as calculated according to equation (8.2). The well developed maximum is clearly seen, which indicates a pattern motion of more than 0.2mm. Measured wall gradients in the flow tube as a function of the Reynolds number are shown in Fig. 8.4. Note the large gradients that can be measured with this technique. However, the largest values are obviously not reliable. This is mainly due to two reasons:

- The very large gradients up to 2500 s^{-1} require smaller temporal distances between the pulses than the minimal 2μs, which could be achieved with the above laser system.
- At high Reynolds numbers the scattering volume is larger than the laminar wall layer.

In Fig. 8.5 measured gradient profiles are shown. The values measured for the smallest Reynolds number of 2250 notably deviate from the other ones.

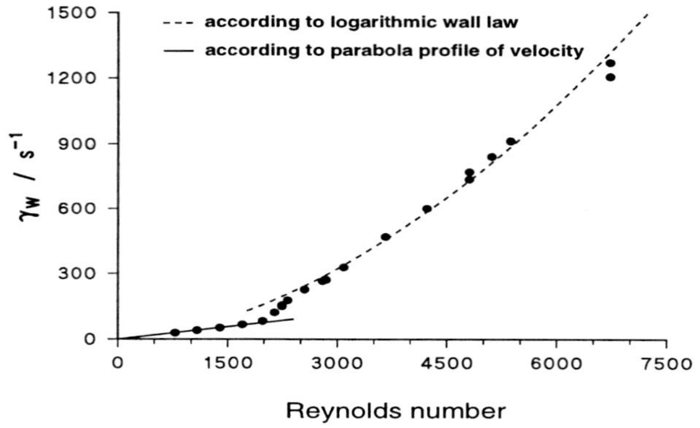

Fig. 8.4. Measured wall gradient as a function of the Reynolds number

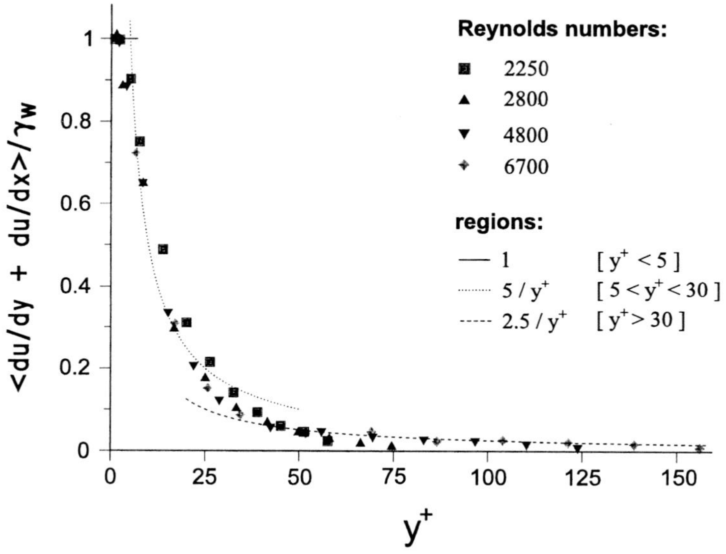

Fig. 8.5. Gradient normalized with the wall gradient as a function of the dimensionless wall distance.

This indicates that in this case the flow turbulence is not yet well developed. Distributions of gradients are shown in Fig. 8.6. In the centre of the flow tube, the distribution turns out to be rather symmetrical, which one would also expect. Near the wall, the distribution becomes definitely asymmetric. We finally show variances of gradients in Fig. 8.7. Here, the transition from laminar to turbulent flow can clearly be seen. Intermittent turbulence already occurs at Reynolds numbers of 1500 - 1600, which is probably caused by pressure fluctuations in the flow system.

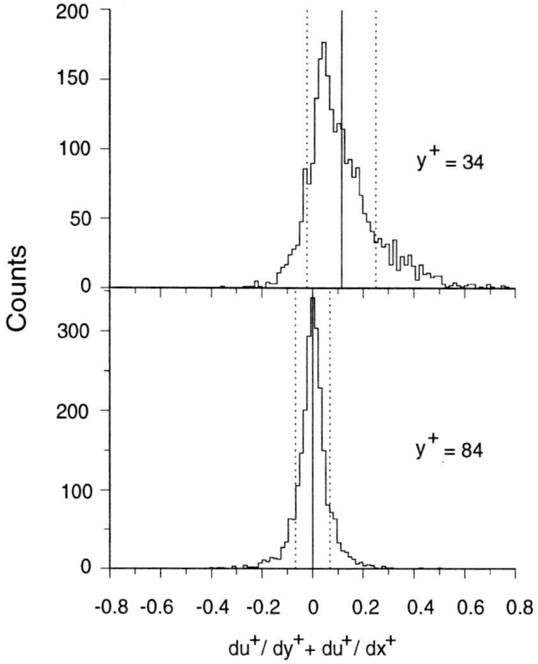

Fig. 8.6. Gradient distributions calculated from 2400 measurements at a Reynolds number of 2800

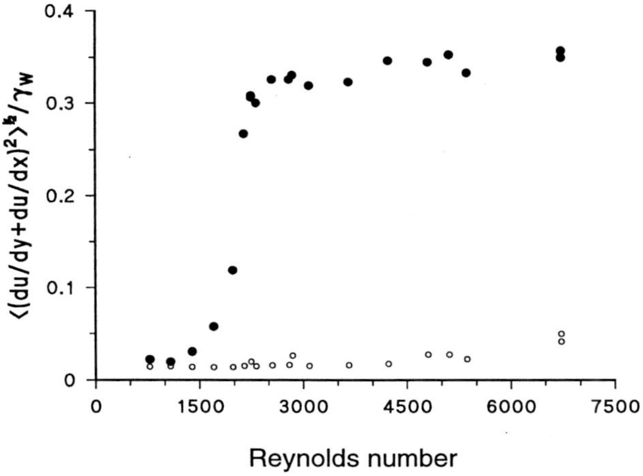

Fig. 8.7. Variations of the gradient component $\Gamma_{xy} + \Gamma_{xx}$ in the near wall region. The open symbols show estimates of the errors produced by the measurement system.

9. DSS: Differential Speckle-Strophometry

9.1 Basic Considerations

9.1.1 Introduction

This measurement technique is based on the property of the intensity of the scattered light expressed by equation (3.62). We will rewrite this equation in the following form:

$$-\langle v_{Px}\rangle = \frac{\left\langle \dot{I}(0,0)\cdot \mathrm{f}(I_x(0,0))\right\rangle}{\langle I_x(0,0)\cdot \mathrm{f}(I_x(0,0))\rangle} \equiv \frac{\langle Z\rangle}{\langle Y\rangle}. \tag{9.1}$$

This defines the quantities Z and Y, which will henceforth be referred to as numerator and denominator, respectively. We are going to call the interesting component of the pattern velocity the 'x-component'. The required averages on the right hand side of this equation will be calculated from the time series of the intensities measured with two detectors. This means that a temporal averaging is performed.

An inherent problem in the technique based on equation (9.1) is the measurement of the derivatives I_x and $\dot{I}$. It is not possible to measure these quantities directly, therefore, one has to resort to a method which allows an approximate experimental determination according to

$$\begin{aligned} \dot{I} &\approx \frac{1}{\tau_S}\left(I\left(0,\frac{\tau_S}{2}\right) - I\left(0,-\frac{\tau_S}{2}\right)\right) \qquad (9.2)\\ I_x &\approx \frac{1}{\delta_D}\left(I\left(\frac{\delta_D}{2}\mathbf{e}_x,0\right) - I\left(-\frac{\delta_D}{2}\mathbf{e}_x,0\right)\right). \end{aligned}$$

The technique described here turns out to be rather complex, which is due to the very large fluctuations of the quantities appearing in this equation. It was therefore necessary to find algorithms which are based on quantities that appreciably reduce the necessary averaging time.

The following subsection will deal with the particulars of this technique, especially with regard to the mentioned problem.

9.1.2 Sample Time and Separation of the Detectors

If the equation (9.2) shall represent a trustworthy approximation to the relation (9.1), one has to keep the quantities δ_D - the separation of the detectors - and τ_S - the sample time - below certain limits. At first sight it seems to be preferable to choose these quantities very small. However, this entails an important disadvantage: The differences of the intensity values occurring in equation (9.2) will then also become very small. Due to the inevitable shot noise in measured intensity values (cf. last paragraph of Sect. 3.1), one then obtains very heavily fluctuating values for these differences. It turns out that these fluctuations are a limiting factor of the technique so it is rather advisable to choose the parameters δ_D and τ_S as big as possible in order to avoid measurement times of many hours. The parameters have to remain small enough, of course, to obtain reliable values for the pattern velocity.

A general relationship between the parameters δ_D and τ_S, and the resulting errors in the determined pattern velocities by use of a definite evaluation algorithm, can only be calculated if the statistical properties of the pattern are known. These are given by the properties of the flow and are therefore not known prior to the measurement.

It is easy to ensure that the approximations expressed by the equations (9.2) are sufficiently accurate if

- $|\delta_D - v_{Px}\tau_S|$ is small compared to the diameter d_x of the coherence area and
- τ_S is small compared to the coherence time τ_C of the pattern.

The quantity d_x - the diameter of the coherence area in the direction in which the pattern motion is measured - can be determined fairly easily, because it follows directly from the dimensions of the q-volume of coherence. The coherence time of the pattern, however, can only be obtained from the temporal auto-correlation function $C_I(\tau)$ of the light intensity, which for that purpose has to be measured. It should be mentioned that the correlation time τ_C of the pattern usually depends very strongly on the scattering geometry. Even if the position of the scattering volume is chosen at a fixed position somewhere in the flow, the correlation times for different scattering geometries may be completely different. This means that it is usually preferable to redetermine the correlation time of the pattern if any quantity describing the scattering geometry is altered.

The inequalities mentioned above are of course only a rough hint as to the choice of the parameters δ_D and τ_S. It is in any case preferable to determine them experimentally. This means that a series of measurements with varying values of δ_D and τ_S has to be performed. Our experiences show that for

$$\begin{aligned} |\delta_D - \boldsymbol{v}_x\tau_S| &\approx 0.1 d_x \\ |\tau_S| &\approx 0.1\tau_C \end{aligned} \tag{9.3}$$

the obtained pattern velocities are reliable.

9.1.3 The Spatial Integration of the Light Intensity

Any light detector integrates the light intensity over the so called detector area. The single point type measurement as supposed by equation (9.2) is principally not possible. Generally, the influence of the detector area on the connection between detector signal and light intensity will be negligible if the detector area is small compared to the coherence area, but in this case the light intensity which hits the detector will be small too. This leads to an increase of the relative signal fluctuations because the influence of the detector noise becomes more important (see also Sect. 9.1.9).

Let us consider the intensity integrated over the detector area

$$\bar{I}(\boldsymbol{r},t) = \int_A I(\boldsymbol{\rho},t)O(\boldsymbol{r}-\boldsymbol{\rho})dA \qquad (9.4)$$

as a new stochastic process. In this relation $\boldsymbol{\rho}$ is the variable location vector in the plane A - the detector area - and $O(\boldsymbol{r})$ the sensitivity function of the detector which quantitatively describes the fact that the sensitivity of the detector depends on where the light hits it.

It is now possible to prove that the velocity of the new pattern $\bar{I}$, as well as the co-ordinate system defined by equation (3.57), remain unaltered if the following condition is met: The function $O(\boldsymbol{r})$ must have mirror symmetry with respect to a reflection at an axis parallel to the direction of differentiation.

The integrated pattern has changed statistical properties as, for instance, an increased coherence area and a larger coherence time. Up to now, only measurements with circular detector areas were performed and we will restrict the following discussions to these. In the experimental set-up the detector areas will be defined by diaphragms in front of the actual detectors, and the combination of detector diaphragm together with the actual detector will in the following be referred to as the detector.

There are three special advantages of circular detector areas:

- Identical diaphragms for two detectors can easily be produced
- Circular diaphragms can easily be adjusted. There are only two parameters which define their respective position: their distance and orientation with respect to each other.
- The co-ordinate system defined by the condition (3.57) is always the same for $I(\boldsymbol{r},t)$ and for $\bar{I}(\boldsymbol{r},t)$.

9.1.4 The Measurement of the Spatial Derivative of the Speckle Pattern

The measurement of the spatial derivative according to equation (9.2) evidently requires two detectors with a respective distance δ_D, which has to

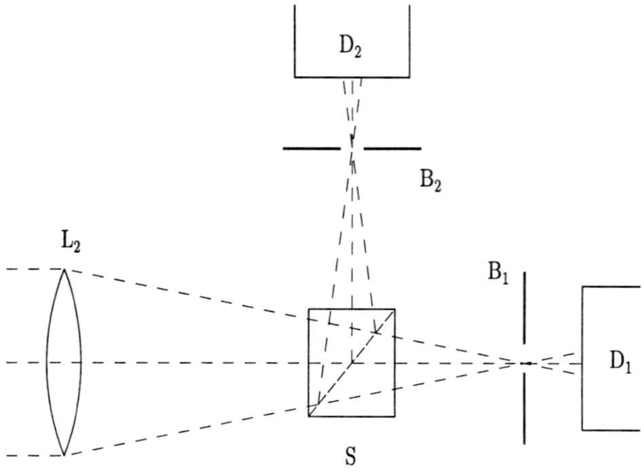

Fig. 9.1. Schematic representation of the optics used for the measurement of the spatial derivative of the speckle pattern. The lens denoted by L_2 corresponds to the one with the same notation in Fig. 5.3. The scattered light passes the beam splitter S and then meets the two detectors D_1 and D_2. The diaphragm B_1 which determines the detector area is slightly displaced perpendicular to the optical axis by an amount δ_D.

be small compared to the diameter d_x of the coherence area. If the detector areas of a size in the order of the coherence area are chosen, an overlap of the two detector areas results. For an experimental realization it is then necessary to use a beam splitter, which produces two identical patterns on two spatially separated observation planes. The two detectors can then be placed in these two planes and be translated independently of each other. This is shown schematically in Fig. 9.1. Even though the intensity of the light is divided by two for each of the detectors, a net gain can be achieved because of the possibility to choose large detector areas.

As to the diameter of the detector area, no quantitative investigations to optimize this parameter have been performed so far.

An important prerequisite for a reliable measurement of the spatial derivative of the integrated pattern is a constant relationship between the two detector sensitivities. This means that the variation of the detector sensitivities across the detectors must be the same:

$$\frac{O_1(\boldsymbol{r})}{O_2(\boldsymbol{r})} = \text{const.} \tag{9.5}$$

Otherwise one would obtain two different signals by the procedure, (according to the right hand side of equation (9.4)), which could no longer serve for a determination of the pattern velocity. No difficulty would of course arise for detectors which have a constant sensitivity across their sensitive area. However, almost all detectors show this variation. Clearly, if the spread in sensitivity is only small or, more generally, the sensitivity variations of the two detectors are fairly equal, one will have no problems. Otherwise, one could resort to the following remedy: The light wave passing through the detector diaphragm has only a very small angle of divergence; it is therefore easy to focus it on a very small spot. If photomultipliers are used, as in our case, the variation of the sensitivity will be rather smooth and change only on a large scale compared to the diameter of the focused light wave. However, if for instance avalanche diodes are used with very small active areas, a very accurate adjustment is required. It is also possible to 'randomize' the light wave with a ground glass or some other means just behind the detector diaphragm. Unfortunately, one will then have a certain amount of loss of the light intensity which is actually seen by the detectors.

If optical elements are positioned somewhere behind the beam splitter one has to make sure that the two light paths remain identical. Lenses with very short focal lengths may especially pose problems; since they have a very large curvature, the amount of reflected light drastically increases with distance from the optical axis. Thus, if two identical lenses are not placed at the same respective positions, the two patterns will not remain identical. It is therefore advisable to use lenses with anti-reflection coating.

The experimental results shown below were obtained with side window photomultipliers and detector diaphragms of < 7 mm diameter. The type of photomultiplier used has some thin wires in front of the photo cathode which leads to the formation of shadows on the cathode. The result can be viewed as a strong variation of the sensitivity of the detector, and the condition (9.5) is not met. Satisfactory measurements were not possible unless an opalescent glass plate and a convex lens were placed between detector diaphragm and photomultiplier. Unfortunately, this reduced the light intensity considerably and increased the signal fluctuations and consequently, the measurement times.

The signal differences of the two detectors are usually not zero even if the light intensities are definitely the same, due to different overall sensitivities. Therefore, one cannot simply interpret the signal differences as proportional to the intensity differences. A fairly easy means to adjust the signals is to reduce the light intensity in the light path of the more sensitive detector. We did that by use of a lambda half plate, which was placed somewhere in the light path in front of the beam splitter. Since the intensity ratio of the two emerging light waves depends, to a certain extent, on their polarization, one can change the intensity ratio by a rotation of this plate.

It is also possible to account for the different average signal of the detectors by special evaluation algorithms. We will discuss this possibility Sect. 9.2.

9.1.5 The Temporal Integration of the Speckle Pattern

The impossibility to measure light intensities at a definite point in space is reflected in the time domain: No detector can measure the intensity at a definite moment in time. In order to obtain high signals with relatively little noise, it is recommended to choose the integration time - usually called sample time - as long as possible. Contrary to the case of spatial integration, where overlapping areas turn out to have advantages, no such advantage can be observed in the time domain. We therefore chose the sample time length so that successive sampling intervals are separated by a short dead time only. This dead time turns out to be necessary because of a peculiarity of any time resolving measurement system: The existence of a correlation time of the system, τ_{Det}. In other words, the response of a detector at a definite time t_0 depends on its history during the time interval $[t_o - \tau_{Det}, t_o]$. If, for instance, the system has detected a photon during this time interval, then the probability to detect another one at the time t is different from that if no such detection occurred.

It is therefore of utmost importance to separate two adjacent sample times by a dead time of the duration $\tau_T > \tau_{Det}$. This means that the detectors are 'switched off' during this time; electric signals delivered by the detectors during that time must not reach the evaluating electronics. This is usually achieved with so called gate electronics. Without this procedure, correlations of the signals of adjacent sample times will arise, that are not based on intensity correlations. The result will be erroneous values for the measured temporal derivatives of the light intensities. In our system the dead time was 1μs.

The above results in a maximum integration time τ_I, which is simply given by

$$\tau_I = \tau_S - \tau_T . \tag{9.6}$$

To obtain small sample times, it is therefore not only important to look for very fast detection systems, but one also has to make sure that these systems have very short correlation times.

The temporal integration can lead to an altered value of the evaluated pattern velocity, different from the case of spatial integration over the detector area. This happens if the speckle boiling is appreciable during one integration time τ_I. So far, neither experimental nor theoretical studies have been performed to asses this effect.

9.1.6 The Size of the Scattering Volume and the Light Power Per Coherence Area

An important quantity for an experimental set-up is the intensity of the light, which hits the detectors. For a comparison of different experimental situations we chose a normalization with the size of the coherence area. For a given power of the illuminating laser, the light intensity integrated over a coherence area is dependent on the size of the scattering volume.

As noted in Sect. 2.3.1 the scattering volume must be smaller than the homogeneity volume, i.e. the volume where equation (4.1) is a valid representation of the velocity field of the fluid.

Lower limits to the size of the scattering volume are determined by the flow velocity: Due to the flow velocity the light scattering particles remain within the scattering volume for a certain time, from which a certain boiling of the pattern results. Clearly, smaller scattering volumes will increase the boiling.

We will now derive an approximation for the light power per coherence area. For that purpose we assume an ellipsoidal q-volume of coherence and consider right angle scattering. The three orthogonal diameters are

- in the direction of the laser beam d_l,
- in the direction of the investigated scattered light d_s,
- perpendicular to the scattering plane d_p.

The optimum concentration of particles according to Sect. 5.3 will be assumed. One then has the result that the fraction $(1 - \exp(-d_l/L))$ of the illuminating laser beam is scattered within the scattering volume.

This is such a concentration that the illuminating laser beam is reduced by one e-th on its entire way of length L through the fluid. The scattering volume will generally be much smaller than the entire light path, which allows us to express the power of the light scattered within the scattering volume by

$$I_S = I_o \cdot \frac{d_l}{L \cdot e}, \tag{9.7}$$

where I_o is the power of the illuminating laser.

The two diameters of the coherence area are determined according to equation (4.57) by the diameters d_l and d_p of the scattering volume. If the dimensions of the scattering volume are large compared to the wavelength of the laser light, one can express the solid angle $\Delta\Omega$, which describes the coherence area as seen from the scattering volume, in the following way

$$\Delta\Omega = \frac{3}{4\pi} \alpha_l \cdot \alpha_p. \tag{9.8}$$

The two angles α_i are those that cover the two diameters of the coherence area as seen from the scattering volume. If these angles are small one can apply the approximation of equation (4.57)

$$\alpha_i \approx \tan(\alpha_i) = \frac{4}{\pi} \cdot \frac{\lambda}{d_i}. \tag{9.9}$$

Let us now assume that the light is scattered isotropically, that means that its intensity is independent of the direction. We then deduce with the help of the last three equations that the whole power of the scattered light $I_{\Delta\Omega}$ into one coherence area is

$$I_{\Delta\Omega} = I_o \cdot \frac{12}{\pi^3 \cdot e} \cdot \frac{\lambda^2}{L \cdot d_p}. \tag{9.10}$$

This important result shows that the light power scattered into one coherence area does only depend on the 'height' of the illuminating laser beam, i.e. on the extension of the scattering volume perpendicular to the scattering plane.

9.1.7 The Shape of the Scattering Volume

We mentioned in the first section of this chapter that the most favourable shape of the scattering volume is a sphere. This shape can be achieved only for right-angle scattering. For this case, the illuminating laser must have a circular cross section. This fixes one diameter of the scattering volume, its depth. The other two diameters have to be shaped by proper use and adjustment of a lens and a circular diaphragm (see Fig. 5.3). It is obvious that the resulting spatial aperture function (see Sect. 4.5) has *no* spherical symmetry, because the laser beam profile is Gaussian, whereas the diaphragm must be described by a step-like function. It is therefore advisable to choose the direction for the spatial differentiation, i.e. the direction of the displacement of the two detectors with respect to each other, within the scattering plane or perpendicular to it. With this choice the aperture function, and thus the coherence area, have the required symmetry (see Sect. 4.7.3).

If only the off-diagonal components of the velocity gradient tensor $\hat{\Gamma}$ are to be measured, which is for instance the case for vorticity measurements, right angle scattering is not necessary, but the scattering angle can be arbitrary. This is discussed in detail in Sect. 4.7.3.

9.1.8 Detectors and Signal Handling

The light detector with the least noise production is the photomultiplier working in the photon count mode. In this mode the detector delivers a short electric pulse for each detected photon, which is the smallest amount of light energy detectable in a light field. The pulse length is usually on the order of 10 ns. The wavelength dependence of the sensitivity of photomultipliers is usually characterized by a range, where it is more or less constant. It decreases sharply for higher and lower wavelengths. The quantum efficiency η is that fraction of photons hitting the sensitive area of the detector that produce an output pulse. From the definition, η is restricted to values given by $0 \leq \eta \leq 1$. The values for commercial photomultipliers are usually in the range of 0.05 to 0.2.

Commercially available systems for photon counting using photomultipliers comprise electronics that amplify the pulses delivered from the photomultiplier, and then pass the signal to a discriminator, which suppresses all pulses below a certain height. By this means one can, to some extent, eliminate the noise produced within the photomultiplier tube and in the electronics up to the discriminator. The level of this discriminator can usually be adjusted to an optimum for any given photomultiplier. After the discriminator, the pulses are usually transformed to normalized TTL-pulses (Transistor-Transistor-Logic).

For the technique described here, the pulses delivered by the electronics must then be counted during one sample time, which gives a measure of the light power falling onto the detector during that time. At the end of the sample time the number of pulses must be transferred to the evaluation electronics. During the dead time the counter must be cleared to be ready for the next sample time. As already stated in Sect. 9.1.5, it is important that during the dead time the light detection process is switched off. The easiest way to achieve this is to suppress the transfer of pulses somewhere along the electronic chain during that time. As evaluation electronics we used a standard PC.

9.1.9 Detector Noise

A light detection system, which works in the photon count mode, always produces a certain amount of noise. Even if a light wave of constant intensity is falling onto the detector, one will observe a fluctuating number of pulses. This behaviour is due to the quantum mechanical nature of the light detection process. It is only the probability to observe n photons within a certain time interval Δt_c for a given light intensity which is determined. This probability is

$$Q(\bar{I}, n) = \exp(-\beta\bar{I})\frac{(\beta\bar{I})^2}{n!} \tag{9.11}$$

where $\bar{I}$ is the light intensity averaged over the considered time interval Δt_c. The quantity β depends on the quantum efficiency of the detector and its area A, and furthermore on the time interval Δt_c and the energy e_ω of a single photon:

$$\beta = \frac{\eta \cdot A \cdot \Delta t_c}{e_\omega} \tag{9.12}$$

Equation (9.11) is valid for ideal detectors, which means that the probability for a photon to be detected at a certain moment is only dependent on the light intensity at that moment, but does not depend on what happened prior to that time.

The function $Q(\bar{I}, n)$ represents a so called Poisson distribution. It is used for the calculation of the average of quantities depending on the light intensity. Consider an arbitrary function $f(\bar{I})$ of the light intensity of which the average is to be calculated. Instead of working with a function of the light intensity, it is preferable to express the light intensity by the number of photons. So we put $f(\bar{I}) = f(m \cdot e_\omega)$ and write the last function simply as $f(m)$. For a constant light intensity we then have

$$\langle f(n) \rangle = \sum_m f(m) Q(\bar{I}, m). \tag{9.13}$$

If the light intensity is not constant, but fluctuates randomly one has to calculate

$$\langle f(n)\rangle = \int_{-\infty}^{\infty} \sum_m f(m)Q(\bar{I},m)P^1(\bar{I})d\bar{I}. \tag{9.14}$$

By use of this equation, the following relations can be calculated:

$$\langle n\rangle = \beta\left\langle \bar{I}\right\rangle \propto \left\langle \bar{I}\right\rangle, \tag{9.15}$$

and

$$\left\langle n^2\right\rangle = \beta^2\left\langle \bar{I}^2\right\rangle + \beta\left\langle \bar{I}\right\rangle \not\propto \left\langle \bar{I}^2\right\rangle. \tag{9.16}$$

It is worth noting here that one generally has

$$\left\langle n^j\right\rangle \not\propto \left\langle (\bar{I})^j\right\rangle \text{ for } j \neq 1. \tag{9.17}$$

We will finally consider the calculation of the average of a quantity depending on the measured photon numbers in two different sample times. We consider an arbitrary function $f(n(i), n(i+j))$ of the two measurements. The average value is then given by

$$\langle f(n(i), n(i+j))\rangle = \int_{-\infty}^{\infty}\int_{-\infty}^{\infty} \sum_{n,m} Q(\bar{I}(i),n)Q(\bar{I}(i+j),m)f(n,m) \cdot P^2(\bar{I},j;\bar{J})d\bar{I}d\bar{J} \tag{9.18}$$

where $\bar{I}(i)$ and $\bar{I}(i+j)$ are the light powers within the respective sample times, and $P^2(\bar{I},j;\bar{J})$ is the corresponding joint probability density. If the sum over n and m can be performed analytically one obtains, by use of the notation

$$\sum_{n,m} Q(\bar{I}(i),n)Q(\bar{I}(i+j),m)f(n,m) = F(\bar{I}(i),\bar{I}(j)),$$

the result

$$\langle f(n(i), n(i+j))\rangle = \int_{-\infty}^{\infty}\int_{-\infty}^{\infty} F(\bar{I},\bar{J})P^2(\bar{I},j;\bar{J})dIdJ. \tag{9.19}$$

Generally, the functions F and f are *different.* Only if f depends linearly on $\bar{I}$ and $\bar{J}$, are the two functions identical - up to a constant factor.

9.2 Two Evaluation Algorithms

9.2.1 General Requirements for the Evaluation Algorithm

Because of the relationships discussed in Sect. 9.1.9, the construction of an evaluation algorithm has to take into account the power of the count numbers which occur in it. If the count numbers occur in the first power only, no bias problems produced by shot noise arise. This means shot noise has no influence on the average values, although increasing shot noise reduces the reliability of experimentally determined average values. If higher powers of count numbers occur in the algorithm, bias effects do arise, i.e. the average values are influenced by the shot noise.

Even with this restriction, the relationships of equation (9.2) allow for a large number of expressions, which can be used to determine the pattern velocity. So it is possible to approximate the temporal and spatial intensity differences occurring in the numerator of equation (9.1) in different ways from count numbers measured at different times: One could calculate the temporal difference with count numbers which were measured prior to those used to calculate the spatial difference or vice versa. Furthermore, the signal is supposed to be stationary, therefore, in any expression only time differences are of importance; different terms in a sum may be shifted arbitrarily on the time scale as long as time differences in a term remain unchanged.

This freedom allows us to formulate some meaningful requirements for an evaluation algorithm.

- The average values of numerator and denominator in equation (9.1) should be formed in such a manner that sample time and detector distance can by chosen as large as possible without the occurrence of systematic errors.
- An expression for the calculation of numerator and denominator should be composed of as little count numbers as possible, and should furthermore have a simple form. This enables fast calculation with a computer. The aim is to do the calculations in real time, so that data acquisition and evaluation can be done synchronously.
- Signal noise, which is always present, must not corrupt the validity of the applied algorithm. This requirement is met if in all expressions the count numbers occur in the first power. If that is not the case one has to investigate whether the average value is actually influenced by noise.
- The so called dark current, i.e. pulses of the detector in absence of light, should not lead to a bias in the evaluated values.
- Different detector sensitivities should have no influence on the evaluated pattern velocity, even if it is fairly easy to adjust the average detector signals on the optical side of the set-up.

It is rather difficult if not altogether impossible to meet all of these five requirements, but it is important not to leave them out of sight when thinking of new algorithms.

In the following parts we denote the count numbers of the two detectors by $n_1(i)$ and $n_2(i)$, where i denotes the i-th sample time.

9.2.2 The Weight Function $f_w(I)$

Although count numbers are measured in this method, it is advisable to first choose a function $f_w(I)$ and then use the function $f_w(n)$ for the evaluation algorithm, because it is generally much easier to quantify the distribution of the measured data without considering shot noise. In contrast, the theoretical treatment including the shot noise usually requires simulations with a PC, which can be very time consuming.

Up to now measurements with two different weight functions $f_w(I)$ have been performed:

1. The C-method using $f_w(I') = I'$
   and
2. the S-method using $f_w(I') = \text{sign}(I')$.

Here, sign (x) is the so called signum function. It is defined by the following properties:

$$\text{sign}(x) = \begin{cases} 1 & \text{for } x > 0 \\ 0 & \text{for } x = 0 \\ -1 & \text{for } x < 0 \end{cases} \tag{9.20}$$

The different effects of this function on the fluctuations of numerator and denominator in equation (9.1) are demonstrated in Fig. 9.2, which shows the time series of intensity values generated by a computer.

It can clearly be seen that for the C-method the average values depend on a few, but very large, intensity values. During many time intervals the values are practically zero and therefore contribute very little to the average. The maximum values in the S-method are noticeably smaller; thus almost all parts contribute to the average. From that it is not difficult to deduce that the fluctuations caused by the algorithm are much smaller in the S-method than in the C-method. It turns out that this is especially true for the respective denominators.

The difference between the two methods will be even more pronounced if gradient correlations are to be measured (see Chap. 10). In that case the numerator of the evaluation algorithm is the product of the two numerators $\dot{I}_1 f_w(I_1')$ and $\dot{I}_2 f_w(I_2')$. The intensities I_1 and I_2 in these two expressions are entirely uncorrelated. If now these two numerators are calculated according to the C-method, one will obtain almost always a very small value. Only in rare occasions two maxima will occur at the same time, which results in a huge contribution to the average value of the combined numerator, which means that this value is almost exclusively determined by these rare events. It is obvious that this mechanism entails enormous fluctuations. The tendency of this effect is also present in the S-method, but much less pronounced.

We will mention here our experience with the application of the two methods. The difference of the two methods for the measurement of average values of the velocity gradient turned out to be not significant. However, velocity gradient correlations could be measured with the C-method in very special

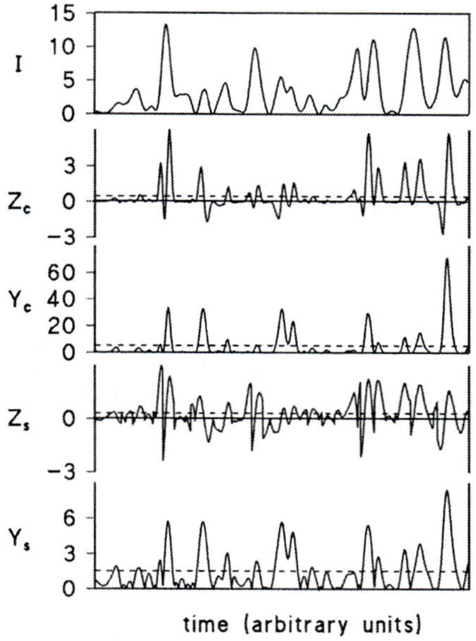

Fig. 9.2. Simulated time series of the intensity I and calculated series of numerator and denominator (c.f. equation (9.1) according to the two methods. The index c denotes the C-method, the index s the S-method. The respective average values are indicated by the dashed lines.

cases only, in particular when the speckle boiling is small. In other cases the average values fluctuate strongly, even after measurement times of a few hours.

The question if there are other functions $f_w(I)$, which are more suitable than the one used in the S-method, has not yet been investigated.

9.2.3 C-Method Algorithm

Numerator and denominator of the equation (9.1) are approximated in this method according to [23, 24] by the following expressions

$$Z_C = \dot{n} \cdot n' \approx \frac{1}{2\tau_S \cdot \delta_D}\{n_1(i)n_2(i+1) - n_1(i+1)n_2(i)\}. \tag{9.21}$$

and

$$Y_C = n' \cdot n' \approx -\frac{1}{\delta_D^2}\{n_1(i+1)n_1(i) + n_2(i+1)n_2(i) - n_1(i+1)n_2(i) \\ -n_1(i)n_2(i+1)\}. \tag{9.22}$$

The first three requirements (see Sect. 9.2.1) are met by these expressions, the other two are not. One can derive, however, an alternative expression for the denominator:

$$\tilde{Y}_C = n' \cdot n' \approx -\frac{1}{\delta_D^2}\left\{\sqrt{\langle n_1(i+1)n_1(i)\rangle\langle n_2(i+1)n_2(i)\rangle} - \langle n_1(i+1)n_2(i)\rangle - \langle n_2(i+1)n_1(i)\rangle\right\}. \quad (9.23)$$

If this expression is introduced in equation (9.1), instead of (9.22), one has an expression where the average does not depend on differences in the detector efficiency. The forth requirement, however, is not fulfilled, i.e. the dark current must be sufficiently small.

9.2.4 S-Method Algorithm

The S-method was presented in [26]. For the S-method the following expressions have been applied:

$$Z_S = \dot{n} \cdot \mathrm{sign}(n') \approx \frac{1}{4\tau_S}\{n_1(i+1) - n_1(i-1) + n_2(i+1) - n_2(i-1)\} \cdot \mathrm{sign}\,(n_1(i) - n_2(i))\,. \quad (9.24)$$

and

$$Y_S = n' \cdot \mathrm{sign}(n') \approx \frac{1}{2\delta_D}\,(n_1(i+1) - n_2(i+1) + n_1(i-1) - n_2(i-1)) \cdot \mathrm{sign}\,(n_1(i) - n_2(i)) \quad (9.25)$$

There are six different count numbers which have to be combined. Fortunately, the calculation of the signum function is a very simple operation, so the calculation of these expressions by a PC is not very time consuming.

The signum function is not linear in its argument, which causes a strong dependence on signal noise. This, however, happens for numerator and denominator in exactly the same way. Consequently, the expression (9.1) is completely independent of signal noise.

As to the last two requirements of Sect. 9.2.1, it is not difficult to show that both are fulfilled. The first factor of the expression for the denominator, however, clearly depends on differences in the detector efficiencies. It is possible to correct for this in the following way:

The sums of the following expressions are done separately:

$$Z_{S1} = \frac{1}{4\tau_S}\{n_1(i+1) - n_1(i-1)\}\mathrm{sign}(n_1(i) - n_2(i)) \quad (9.26)$$

$$\begin{aligned}
Z_{S2} &= \frac{1}{4\tau_S}\{n_2(i+1) - n_2(i-1)\}\mathrm{sign}(n_1(i) - n_2(i)) \\
Y_{S1} &= \frac{1}{2\delta_D}\{n_1(i+1) + n_1(i-1)\}\mathrm{sign}(n_1(i) - n_2(i)) \\
Y_{S2} &= \frac{1}{2\delta_D}\{-n_2(i+1) - n_2(i-1)\}\mathrm{sign}(n_1(i) - n_2(i))
\end{aligned} \quad (9.27)$$

and in addition

$$M_{S1} = n_1(i) \tag{9.28}$$
$$M_{S2} = n_2(i)$$
$$\tag{9.29}$$

is calculated. From the last two equations the ratio of detector sensitivities can be obtained:

$$m_{12} = \frac{\langle M_{S1}\rangle}{\langle M_{S2}\rangle}. \tag{9.30}$$

The expression

$$\frac{\langle Z_{S1}\rangle + m_{12}^2 \langle Z_{S2}\rangle}{\langle Y_{S1}\rangle + m_{12}^2 \langle Y_{S2}\rangle} = \langle v\rangle \tag{9.31}$$

yields the pattern velocity and does not depend on differences of the detector sensitivities. The dark signals only have an influence on the value of the calculated pattern velocity if they are different for the two detectors, because in that case different additional terms appear in numerator and denominator on the right hand side of equation (9.26). However, it is also possible to correct for that effect. One just has to measure the dark signals with the illuminating laser switched off.

9.2.5 Limits of the Method

With the set-up we used in our experiments the maximum velocity gradient which can reliably be measured is about 100 s^{-1}, which is not very large. The size of this value can be estimated in the following way: The experiments show that during the sample time τ_S the pattern should not move a distance larger than about one tenth of the diameter d_C of the coherence area. This diameter at the distance L from the scattering volume - we consider a set-up with no optical elements in the detection channel - is given by equation (4.57). With a velocity gradient element γ within the scattering volume which causes pure speckle motion, the velocity v_P of the pattern in the distance L is obtained with the help of equations (4.72) and (4.73):

$$v_P = L \cdot |\omega| = L \cdot \left[\hat{\Gamma}^T \cdot (\boldsymbol{e}_i - \boldsymbol{e}_o)\right]_\perp \approx L\gamma. \tag{9.32}$$

Now the time Δt which the pattern needs to travel one tenth of the speckle size is given by:

$$\Delta t = \frac{0.1 d_C}{v_P} \approx \frac{4L\lambda}{10\pi d_S}\frac{1}{L\gamma} \approx \frac{\lambda}{10 d_S \gamma}. \tag{9.33}$$

With a diameter of the scattering volume of 100 μm and a wavelength of the laser light of $\approx 0.5\mu m$ we get

$$\Delta t \approx \frac{5 \cdot 10^{-4}}{\gamma}. \tag{9.34}$$

The sample time should not be larger than this value Δt. With photomultipliers the minumum sample time is in the order of a few micro seconds, otherwise the number of detected photons in one sample time is too low. With $\tau_S = 5\ \mu$s one then obtains for the maximum velocity gradient a value of about 100 s^{-1}.

There is a possibility to appreciably increase this value by increasing the effective size of the coherence area. We already mentioned in Sect. 9.1.3 that spatial integration of the pattern caused by a detector diaphragm leads to a larger effective coherence area. However, using a diaphragm with sharp edges does not decrease the slope of the effective pattern below a certain value determined by the speckle size. As a consequence, for the measurement of the spatial and temporal derivatives of the pattern, one can not simply increase the distance of the diaphragms or increase the sample time. A smooth pattern with a size of the coherence area much larger than the speckle size can be obtained with diaphragms in front of the detectors with sufficiently smoothly varying transmission. It is then possible, for a given pattern velocity, to place the detectors further apart and choose longer sample times. Consequently, it is then possible to measure larger velocity gradients. A further advantage of using large diaphragms of this type results from the increased light intensity on the detectors: The influence of the shot noise will be reduced.

9.3 Experiments and Results

In this chapter we present results obtained with the C-method only, since those obtained with the S-method are not different from them. The S-method was developed to measure gradient correlations, and the results obtained with this method will be given in the next chapter.

9.3.1 Test of the Set-Up

The fairly high demands for the proper optical alignment with this method render test measurements almost inevitable. Not only measurements on a 'test flow' as described in Sect. 5.6 are advisable, but also tests of numerator and denominator, as given in equations (9.21) and (9.22), as a function of the detector distance are recommended. For small detector distances δ_D the numerator must depend linearly on δ_D and it must have the value zero at $\delta_D = 0$. The denominator must show a pure square dependence and must also have the value zero at $\delta_D = 0$.

The form of the denominator must be independent of the size of the pattern velocity, because it depends only on the spatial form of the pattern. However, speckle boiling does effect the size of it. The slope of the numerator must turn out to be proportional to the velocity gradient, it depends in the same way as the denominator on the speckle boiling.

The first reliable results with the technique described in this chapter were obtained by use of the C-method. Water within a rotating cuvette mounted on a commercial turntable served as the flow. By this means a constant known velocity gradient was guaranteed, and three different gradients could be chosen. It is also possible to change the pattern boiling by varying the position of the scattering volume within the cuvette: At the rotation axis, there is no boiling caused by the overall velocity, and that type of boiling increases linearly with the distance from the axis.

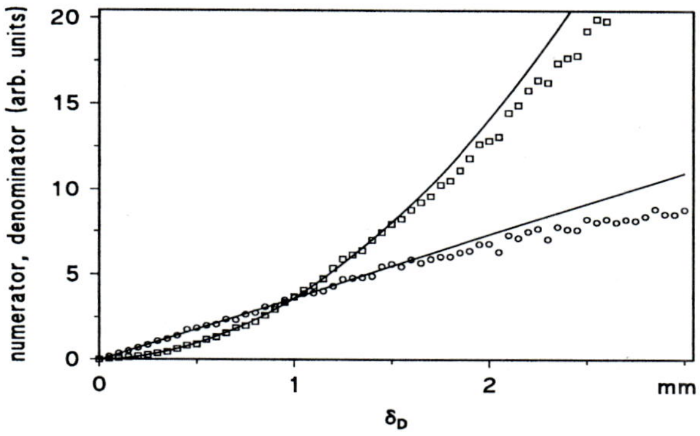

Fig. 9.3. Measured values of the numerator (equation (9.21)) and the denominator (equation (9.22)) as functions of the detector distance. The smooth curves show the linear and quadratic behaviour of numerator and denominator, respectively.

These quantities were measured with an experimental set-up which is essentially the same as the one shown in Fig. 5.3. The optical set-up led to a diameter of the coherence area at the detectors of $\approx$ 20mm. Figure 9.3 shows the measured numerator and denominator as functions of the detector distance. Beginning at a distance of about 1.5 mm, deviations from the linear and quadratic behaviour can be observed. Nevertheless, the ratio of the two quantities remains constant even for larger detector distances. An estimation on the basis of equation (9.2) requires the knowledge of the sample time, which for these measurements was 128 μs. We then have $v \cdot \tau_S \approx 0.5$mm.

In Fig. 9.4 the dependence of the determined pattern velocity as a function of the horizontal detector distance is plotted.

The raw data, which were obtained by simply dividing numerator by denominator, show a strong deviation for small detector distances. This is presumably caused by a slight misalignment of the detectors: If there is a small vertical shift of one detector with respect to the other, this leads to a rotation of the effective direction of the line combining the two detectors. This means that for small distances the effective direction, in which the spa-

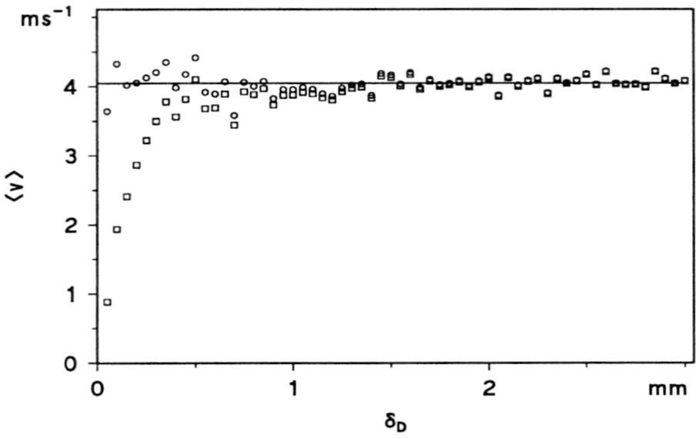

Fig. 9.4. The dependence of the determined pattern velocity on the distance δ_D of the two detectors. The squares show the measured values, the circles are values obtained from these after a correction (for particulars see text).

tial derivative of the pattern is determined, is also rotated. If the data are corrected with the assumption of a vertical distance of 0.2 mm, one obtains the data denoted in the figure by circles.

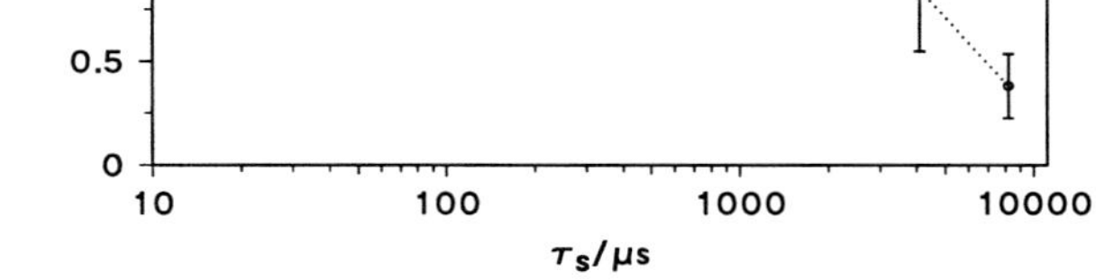

Fig. 9.5. The dependence of the measured pattern velocity on the sample time τ_S (average values and standard deviations determined from ten single measurements).

The influence of the sample time on the determined pattern velocity is shown in Fig. 9.5. A clearly noticeable deviation from the nominal pattern velocity occurs for sample times longer than 1 ms. In this case we have $v \cdot \tau_S =$ 2 mm. The detector distance was 1.5 mm.

9.3.2 Results of Flow Measurements

Figure 9.6 shows measured wall gradients in the flow channel described in Sect. 7.2.1. The distance of the scattering volume from the wall was 0.3 mm.

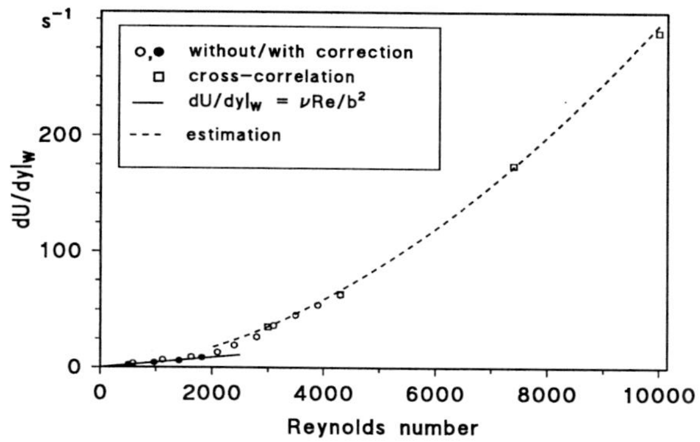

Fig. 9.6. Mean gradient du/dy in the distance y = 0.3 mm from the channel wall

The corrections for the laminar flow (Reynolds numbers smaller than about 2000) were derived from measurements of the velocity profile of the channel. These showed that the flow was not symmetric with respect to the channel centre. Furthermore, the figure incorporates values obtained with the cross correlation method. In the laminar regime there is a linear relation between velocity gradient and Reynolds number, while in the turbulent regime the gradient depends on a higher power of the Reynolds number. The dashed curve represents a fit of the logarithmic law of the wall (see for instance [27]) to the data in the turbulent regime, which should be valid almost up to the middle of the channel.

Normalized gradient profiles as measured with the method described in this chapter are shown in Fig. 9.7. Unfortunately, values for Reynolds numbers higher than >4300 could not be measured, since the shortest sample time, which could be chosen, was 5 μs with the used electronics. The dashed curve for normalized wall distances larger than 5 was again calculated according to the logarithmic law of the wall.

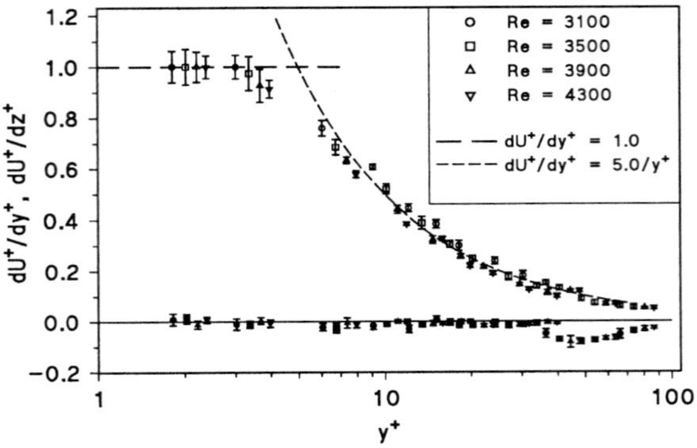

Fig. 9.7. Mean normalized gradient du^+/dy^+ (empty symbols) and $\mathrm{du}^+/\mathrm{dz}^+$ (full symbols) as a function of the normalized wall distance y^+

10. Measurement of Velocity Gradient Correlations

10.1 Basic Considerations

10.1.1 Introduction

On the basis of the method described in the last chapter, it is possible to measure velocity gradient correlations. Let us first treat the theoretical basis for the extension of the method outlined in the last chapter. It seems obvious to check whether the following straightforward generalization of equation (9.1) is valid:

$$\frac{\langle Z_1 \cdot Z_2 \rangle}{\langle Y_1 \cdot Y_2 \rangle} = \langle v_1 \cdot v_2 \rangle \,, \tag{10.1}$$

The two indices refer to two scattering experiments which are performed at the same time in the flow. In this chapter we will consider the case where the scattering volumes are identical, i.e. correlations of the same or different velocity gradients at the same location will be considered. An analysis reveals that this equation is indeed true provided that the light intensities measured in experiment 1 are entirely independent of those measured in experiment 2. This may also be expressed in the following way: A correlation between Z_1 and Z_2 must exclusively be caused by the velocity of the two patterns of scattered light. If this condition is fulfilled then the denominators Y_1 and Y_2 are uncorrelated, since they do not depend on the speckle velocity.

There are two different effects that can cause the intensities of the scattered light to be correlated: Firstly, the number of scattering particles within the scattering volume determines the intensity of the scattered light. Since the two scattering experiments are done with the same scattering volume, the fluctuations of the number of scattering particles within it is the same for both experiments, and therefore a certain correlation of the two light intensities is obvious. Fortunately, denominator and numerator in the equation (10.1) fluctuate in the same way. Therefore, no net effect due to these fluctuations occurs as far as the velocity gradient correlations are concerned.

Secondly, the scattered light originating in the same scattering volume is correlated whenever the scattering vectors lie in the same q-volume of coherence. In this case, however, the fluctuations in numerator and denominator on the right hand side of equation (10.1) do not cancel. For this reason, measurements with the same scattering volume are only possible if the two $\boldsymbol{q}$-vectors are sufficiently different, so that the two q-volumes of coherence are so far apart that no correlations will result. This is the case if their distance is at least five diameters of the q-volumes of coherence.

Measurements on the basis of the relation (10.1) showed that the fluctuations of the products $Z_1 \cdot Z_2$ and $Y_1 \cdot Y_2$ are so big that no reliable correlation values could be obtained. It is only with some modifications of this equation, that the intended measurements can be performed.

We mentioned at the end of Sect. 3.5 that the denominators are not dependent on the pattern velocity. It is therefore possible to change the averaging procedure given in the denominator of equation (10.1) to separate averaging, i.e. replace this denominator by $\langle Y_1 \rangle \langle Y_2 \rangle$. The result is a much faster process of convergence of the combined denominator than that of equation (10.1).

A very helpful alteration of the numerator is based on the fact that the gradient at a location in the flow remains constant for relatively long times compared to the correlation time of the light intensity in the observation plane. In our case, (see below), the correlation times τ_Γ of the velocity gradients were in most cases longer than 10 ms, whereas the sample times were usually shorter than 20 μs. It is therefore admissible to perform the averaging procedure for the numerator in two steps: A first averaging is performed with the two factors of the numerator *separately* over a so called intermediate averaging time T_Z, which is subject to the condition $T_Z << \tau_\Gamma$. Since the pattern velocities can be considered constant during that time, and these two factors can only be correlated through correlations of the corresponding pattern velocities, the product of these two intermediate averaged factors can be considered as 'instantaneous' values as far as the pattern velocity is concerned. The final long time averaging of this product therefore leads to the same result as a straightforward averaging of the numerator in equation (10.1). We can thus write:

$$\langle Z_1 \cdot Z_2 \rangle = \left\langle \overline{Z_1^T} \cdot \overline{Z_2^T} \right\rangle , \tag{10.2}$$

where the overline denotes the intermediate averaging. The big advantage of the intermediate averaging is that the resulting factors have a much smaller variance, therefore the convergence of the right hand is much faster than that of the left hand side, which saves an enormous amount of measurement time.

If we combine these two results we finally have

$$\frac{\left\langle \overline{Z_1^T} \cdot \overline{Z_2^T} \right\rangle}{\langle Y_1 \rangle \cdot \langle Y_2 \rangle} = \langle v_1 \cdot v_2 \rangle , \tag{10.3}$$

Measurements showed that acceptable measurement times (on the order of 30 min) could only be achieved with this variant of the S-method. Measurements with the C-method showed even after a few hours no sufficiently stable result.

With this method, the temporal correlation function can also be measured. We are going to write down in its digital form:

$$C_{12}(i) = \frac{\frac{1}{N}\sum_j \left\langle \overline{Z^T}_{1,j} \cdot \overline{Z^T}_{2,j+i} \right\rangle}{\frac{1}{N}\sum_j \left\langle \overline{Y^T}_{1,j} \right\rangle \cdot \frac{1}{N}\sum_j \left\langle \overline{Y^T}_{2,j} \right\rangle}. \tag{10.4}$$

The indices i and j in this equation refer to the intermediate averaging time T. Consequently, the argument i appearing in $C_{12}(i)$ denotes the time $i \cdot T$.

This longer time scale T is much better suited for the representation of the correlation function than the sample time, since the velocity gradient is essentially constant over very many sample times.

10.1.2 Symmetry Considerations

The number of independent correlations between the components of the velocity gradient tensor can reduce considerably if the flow under consideration has certain symmetries. Of course, in a turbulent flow the notion of symmetry refers to properties of average values. In an arbitrary flow, the velocity gradient tensor has 9 independent components, from which follows a number of 81 correlations. However, products which differ only in the succession of the factors in the cross-correlation terms are identical. Therefore, from the 72 cross-correlations only 36 remain, and altogether we are left with 45 correlations.

Incompressibility. In incompressible fluids the trace of the velocity gradient tensor is zero (see equation 4.4), which reduces the number of independent components of the velocity gradient tensor to 8. One then has 8 independent auto-correlations and 28 cross-correlations.

Mirror Symmetry. The flows investigated with the technique described here are a channel and a tube flow, which both have a large height-to-width ratio, so it is a good approximation to consider them as systems with the x-y-plane as a mirror plane. In the open channel flow this plane coincides with the surface of the fluid, in the rectangular tube with the plane in the middle of the tube. Consequently, all average values must be invariant under the transformation $z \to -z$. With this transformation, also the z-component of the velocity is reversed according to $u_z \to -u_z$, and again, all average values must remain the same under this transformation.

Let us now calculate how this symmetry property changes the number of independent velocity gradient components. To this end, we first consider the behaviour of the gradient components. A general element of the gradient tensor can be written in the form Γ_{ij}, where i and i are elements of the set $\{x, y, z\}$. We then have

$$\Gamma_{ij} = \frac{du_i}{dx_j}. \tag{10.5}$$

The mentioned transformation does only change the sign of an element of Γ_{ij}, which has exactly one of its two indices equal to z. It is obvious, that for instance Γ_{xy} is not affected, but Γ_{zz} also remains unaltered because the transformation causes a double reversion of the sign. Thus, if $\langle \Gamma_{ij} \cdot \Gamma_{lm} \rangle$ contains the index z an odd number of times, it changes sign under the mentioned transformation, otherwise no change occurs. Due to the symmetry of the flow, we now know that all gradient correlations which change sign must vanish, therefore 16 cross correlations are identical to zero, and only 8 auto- and 16 cross-correlations remain.

For a rotationally symmetric flow as, for instance, a circular jet, one obtains an analogous result. The co-ordinates suited to the description of this flow are the so called cylindrical co-ordinates r, z and ϕ, where the z-axis points along the centre of the flow, r denotes the distance of a point from this axis and ϕ denotes the polar angle in the plane perpendicular to the z-axis. In this case the flow has the symmetry property that it is invariant under the transformation $\phi \rightarrow -\phi$. This symmetry leads to the same number of independent correlations as for the case discussed first.

Translational Symmetry. A further possible symmetry in flows is a translational symmetry, which means that the properties of the flow remain unchanged if one moves to another location in one or more directions. Let us consider the flows mentioned above with a translational symmetry in the x-direction. It is then obvious that all average quantities must be independent of x, which implies that the derivative of these quantities with respect to x disappears. This property, together with the mirror symmetry considered in the last subsection, lets us derive two relationships between gradient components as follows:

According to the translational symmetry we have:

$$\frac{d}{dx}\left\langle \frac{du}{dz} \cdot w \right\rangle = \left\langle \frac{d^2u}{dxdz} \cdot w \right\rangle + \left\langle \frac{du}{dz} \cdot \frac{dw}{dx} \right\rangle = 0. \tag{10.6}$$

(Note that sequence of the operations 'averaging' and 'differentiation' is entirely arbitrary.) From the mirror symmetry we deduce

$$\left\langle \frac{du}{dx} \cdot w \right\rangle = 0 \tag{10.7}$$

because w reverses sign by a reflection at the x-y-plane. Differentiating the last equation with respect to z yields

$$\left\langle \frac{d^2u}{dxdz} \cdot w \right\rangle + \left\langle \frac{du}{dx} \cdot \frac{dw}{dz} \right\rangle = 0. \tag{10.8}$$

Combining equations (10.6) and (10.8) leads to

$$\left\langle \frac{du}{dx} \cdot \frac{dw}{dz} \right\rangle = \left\langle \frac{du}{dz} \cdot \frac{dw}{dx} \right\rangle . \tag{10.9}$$

In a completely analogous way the following equation is obtained:

$$\left\langle \frac{dv}{dx} \cdot \frac{dw}{dz} \right\rangle = \left\langle \frac{dv}{dz} \cdot \frac{dw}{dx} \right\rangle . \qquad (10.10)$$

Form the above equations we finally conclude that for a channel flow with the considered symmetries only 8 auto- and 10 cross-correlations need to be measured. Of course, these are special correlations of definite components of the velocity gradient or linear combinations of them. If one is going to measure *all* non-vanishing correlations with a certain experimental set-up, one has to ensure that really all of the 18 correlations can be calculated from those *measured*. Generally, it is advisable to measure more than 18 correlations, so that one has the possibility to check the consistency of the measured values. It is also possible to test to what extent the assumed symmetry properties hold.

10.1.3 Experimental Considerations

The complexity of the experimental set-up depends a good deal on the amount and the type of information to be obtained. With the most simple set-up, as depicted in Fig. 5.3, and the technique presented in Sect. 10.1 not only gradients, but also auto-correlation functions can be obtained. However, care must be taken with the zeroth channel of this correlation function: The two intermediate averaged factors are identical, and they are therefore correlated through their common fluctuations resulting from shot noise and the varying intensity. To a lesser extent this unwanted correlation is still present in the first channel of the correlation function, caused by an intensity correlation in the sample times around the limits of the intermediate averaging periods. If, however, the intermediate averaging is performed over a large number of sample times this correlation is negligibly small.

For the measurement of cross-correlations a set-up with at least two measurement channels must be used. Generally, the optimum sample times and intermediate averaging times for the two channels are different. So it is important that these times can be chosen separately in the electronics of the two channels. Obviously, the longer intermediate averaging time must be a multiple of the smaller one.

To measure more gradient correlations simultaneously, not only the number of measurement channels can be increased, but it is also possible to use two illuminating light beams. With two illuminating light beams and at least two measurement channels, the restriction to right angle scattering can be maintained only if the two illuminating light beams have exactly opposite directions.

The two light beams must, of course, not illuminate the scattering volume at the same time, since then two patterns with two different velocities would exist in the detection plane. So rather switching the illuminating light beam between two directions is required instead of using two light beams. If this switching is done twice in an intermediate averaging time - preferably exactly in the middle and at the end - one is able to measure two gradient components with one measurement channel at the same time, and even the correlation between the two components. Of course, the detector signals for the two beam directions must be well discriminated against each other by synchronizing the switching device - usually a Pockels cell - with the electronic signal switch.

Let us consider more closely the set-up with two opposite illumination directions and two measurement channels. With each of the two measurement channels, four gradient components can be measured, because there are two illumination directions and two orthogonal directions for the measurement of the pattern velocity. So altogether eight linear combinations of gradient components can be obtained. A closer inspection reveals, however, that only seven of them are linearly independent. If the two measurement channels analyse light scattered in exactly opposite directions then there are actually only four independent linear combinations, these are the same ones obtained with only one measurement channel. It is worth noting that in the special case of right angles between the two measurement channels, one obtains *all* non-diagonal elements of the gradient tensor, i.e. all components of the vorticity.

We are now going to consider the gradient correlations. For the set-up just specified, all possibilities will be given explicitly. The following notation will be used: The illumination direction is denoted by the first index $i = 1, 2$. The second index j specifies the measurement channel. With two measurement channels we have $j = 1, 2$. A further index k indicates which of the two orthogonal pattern velocity components in a channel is measured. A particular gradient component - more precisely: the linear combination of gradient components - which is measured in one channel is then completely determined by three indices and we call it $\tilde{\Gamma}_{ijk}$. $\tilde{\Gamma}_{ijk}$ can be expressed by the gradient components Γ_{mn} of the flow if according to Sect. 5.8.1 the following quantities are known:

- the wave vector of the illuminating light beam (index i)
- the wave vector of the measured light component (index j)
- the direction in which the pattern velocity is measured (index k)

It is then possible to measure the following 28 gradient correlations:

Auto correlations	cross-correlations different channels same illumination	cross-correlations different channels different illumination	cross-correltions same channel different illumination
$\langle\tilde{\Gamma}_{111}\cdot\tilde{\Gamma}_{111}\rangle$	$\langle\tilde{\Gamma}_{111}\cdot\tilde{\Gamma}_{121}\rangle$	$\langle\tilde{\Gamma}_{111}\cdot\tilde{\Gamma}_{221}\rangle$	$\langle\tilde{\Gamma}_{111}\cdot\tilde{\Gamma}_{211}\rangle$
$\langle\tilde{\Gamma}_{112}\cdot\tilde{\Gamma}_{112}\rangle$	$\langle\tilde{\Gamma}_{111}\cdot\tilde{\Gamma}_{122}\rangle$	$\langle\tilde{\Gamma}_{111}\cdot\tilde{\Gamma}_{222}\rangle$	$\langle\tilde{\Gamma}_{112}\cdot\tilde{\Gamma}_{212}\rangle$
$\langle\tilde{\Gamma}_{121}\cdot\tilde{\Gamma}_{121}\rangle$	$\langle\tilde{\Gamma}_{112}\cdot\tilde{\Gamma}_{121}\rangle$	$\langle\tilde{\Gamma}_{112}\cdot\tilde{\Gamma}_{221}\rangle$	$\langle\tilde{\Gamma}_{121}\cdot\tilde{\Gamma}_{221}\rangle$
$\langle\tilde{\Gamma}_{122}\cdot\tilde{\Gamma}_{122}\rangle$	$\langle\tilde{\Gamma}_{112}\cdot\tilde{\Gamma}_{122}\rangle$	$\langle\tilde{\Gamma}_{112}\cdot\tilde{\Gamma}_{222}\rangle$	$\langle\tilde{\Gamma}_{122}\cdot\tilde{\Gamma}_{222}\rangle$
$\langle\tilde{\Gamma}_{211}\cdot\tilde{\Gamma}_{211}\rangle$	$\langle\tilde{\Gamma}_{211}\cdot\tilde{\Gamma}_{221}\rangle$	$\langle\tilde{\Gamma}_{211}\cdot\tilde{\Gamma}_{121}\rangle$	
$\langle\tilde{\Gamma}_{212}\cdot\tilde{\Gamma}_{212}\rangle$	$\langle\tilde{\Gamma}_{211}\cdot\tilde{\Gamma}_{222}\rangle$	$\langle\tilde{\Gamma}_{211}\cdot\tilde{\Gamma}_{122}\rangle$	
$\langle\tilde{\Gamma}_{221}\cdot\tilde{\Gamma}_{221}\rangle$	$\langle\tilde{\Gamma}_{212}\cdot\tilde{\Gamma}_{221}\rangle$	$\langle\tilde{\Gamma}_{212}\cdot\tilde{\Gamma}_{121}\rangle$	
$\langle\tilde{\Gamma}_{222}\cdot\tilde{\Gamma}_{222}\rangle$	$\langle\tilde{\Gamma}_{212}\cdot\tilde{\Gamma}_{222}\rangle$	$\langle\tilde{\Gamma}_{212}\cdot\tilde{\Gamma}_{122}\rangle$	

For a flow without any symmetries in the sense considered above, these 28 correlations would be entirely independent. Since a channel flow has symmetries, dependencies do exist between these correlations.

A general flow without any symmetry needs two opposite illumination directions and three measurement channels for the determination of all gradient correlations. The measurement channels must be adjustable as mentioned before. This set-up allows the measurement of 12 different auto-correlations and 54 cross-correlation functions. Obviously, these correlation functions cannot be independent, so it is not necessary to measure all of them.

10.1.4 Data Evaluation

If the measured gradient correlations are expressed in the co-ordinate system of the flow, one generally has linear combinations of many terms. For scattering geometries with scattering vectors $\boldsymbol{k}_i - \boldsymbol{k}_o$ in an arbitrary direction, and also arbitrary measured pattern velocity components, these linear combinations will contain almost all gradient correlation functions, which amount to 45. The coefficients are uniquely determined by the scattering geometry in a way analogous to that described in Sect. 5.8.2. We are now going to describe how, from the measured correlation functions and under consideration of the symmetry relations, the Cartesian correlation functions can be obtained.

The number of equations, which follow from the symmetry of the flow, are denoted by N_S, the number of measured correlations N_M. The number of all the gradient correlations is $N_A = 45$. The unknown correlations, which are to be calculated from those measured, are written in form of the vector $\boldsymbol{x}$ with N_A components. Each component of the vector $\boldsymbol{x}$ is then a definite correlation in a proper co-ordinate system, preferably the Cartesian co-ordinate system of the flow. For circular symmetric flows, however, one can also choose a cylindrical co-ordinate system. The set of measured gradient correlations should also be arranged in form of a vector $\boldsymbol{y}$ with N_M components.

The N_S symmetry equations

$$\begin{array}{llllll} S_{11}\cdot x_1 & + S_{12}\cdot x_2 & + S_{13}\cdot x_3 & + \ldots + & S_{1N_A}\cdot x_{N_A} & = 0 \\ S_{21}\cdot x_1 & + S_{22}\cdot x_2 & + S_{23}\cdot x_3 & + \ldots + & S_{2N_a}\cdot x_{N_A} & = 0 \\ S_{31}\cdot x_1 & + S_{32}\cdot x_2 & + S_{33}\cdot x_3 & + \ldots + & S_{3N_A}\cdot x_{N_A} & = 0 \\ \cdot & \cdot & \cdot & \cdot & \cdot & \cdot \\ \cdot & \cdot & \cdot & \cdot & \cdot & \cdot \\ \cdot & \cdot & \cdot & \cdot & \cdot & \cdot \\ S_{N_S 1}\cdot x_1 & + S_{N_S 2}\cdot x_2 & + S_{N_S 3}\cdot x_3 & + \ldots + & S_{N_S N_A}\cdot x_{N_A} & = 0 \end{array} \tag{10.11}$$

will now be written in form of a matrix:

$$\hat{S}\cdot\boldsymbol{x} = \boldsymbol{0}, \tag{10.12}$$

where $\boldsymbol{0}$ is a N_S-dimensional zero vector and $\hat{S}$ a rectangular matrix with N_S rows and N_A columns.

The problem is now to determine the unknown correlations $\boldsymbol{x}$ so that the measured correlations $\boldsymbol{y}$ follow from them.

For that purpose, let us assume that the required correlations were known. In that case one could calculate what the 'ideal' measured values, denoted by $\boldsymbol{y}^I$, would be. And these values would be given by a linear combination of the form

$$\begin{array}{llllll} M_{11}\cdot x_1 & + M_{12}\cdot x_2 & + M_{13}\cdot x_3 & + \ldots + & M_{1N_A}\cdot x_{N_A} & = y_1^I \\ M_{21}\cdot x_1 & + M_{22}\cdot x_2 & + M_{23}\cdot x_3 & + \ldots + & M_{2N_A}\cdot x_{N_A} & = y_2^I \\ M_{31}\cdot x_1 & + M_{32}\cdot x_2 & + M_{33}\cdot x_3 & + \ldots + & M_{3N_A}\cdot x_{N_A} & = y_3^I \\ \cdot & \cdot & \cdot & \cdot & \cdot & \cdot \\ \cdot & \cdot & \cdot & \cdot & \cdot & \cdot \\ \cdot & \cdot & \cdot & \cdot & \cdot & \cdot \\ M_{N_M 1}\cdot x_1 & + M_{N_M 2}\cdot x_2 & + M_{N_M 3}\cdot x_3 & + \ldots + & M_{N_M N_A}\cdot x_{N_A} & = y_{N_M}^I \end{array} \tag{10.13}$$

The M_{ij} are given by the angles of the different scattering geometries which were used for the measurements. Let us recall what the meaning of the indices i and j in the last equation is: i denotes the number of the gradient component. In the example discussed in the last subsection it varies from 1 to 28. The second index j denotes all the different gradient correlations and, therefore, varies from 1 to 45.

The last equation can also be written in a compact form as

$$\hat{M}\cdot\boldsymbol{x} = \boldsymbol{y}^I \tag{10.14}$$

It is obviously next to impossible to obtain the y_i^I in an actual measurement. But one can hope that the measured values y_i do not differ too much from the ideal ones. A measure of the difference of the two sets of data is apparently the quantity

$$Q = \left(\boldsymbol{y}^I - \boldsymbol{y}\right)^2 = \sum_i \left(y_i^I - y_i\right)^2 \tag{10.15}$$

which for a good measurement should be very small. This idea is used to determine the unknown x_i from the measured y_i: Consider the quantity Q as a function of the x_i via the y_i^I (see equation (10.13)). Then search the set of unknown x_i which makes Q a minimum. This set is then 'most likely' the set of the true correlations. (The method described is a 'maximum likelihood method'.)

Let us write down explicitly the quantity Q as a function of the measured correlations y_i and the unknown x_i:

$$Q(\boldsymbol{x}) = \sum_i \left(\sum_j M_{ij} x_i - y_j\right)^2 \tag{10.16}$$

$$= \left(\hat{M} \cdot \boldsymbol{x} - \boldsymbol{y}\right)^2 . \tag{10.17}$$

How the minimum of $Q(\boldsymbol{x})$ under consideration of the symmetry relations 10.11 can be determined depends on how reliable the validity of these relations is considered. If they are considered absolutely valid it is advisable to use them for the reduction of the number of the variables x_i in equation (10.13). This is especially easy in the co-ordinate system adapted to the flow. In the case of a channel flow, for instance, many of the x_i are simply zero. This means that in the system of equations (10.13), the corresponding columns of the matrix $\hat{M}$, as well as the corresponding elements in the vector $\boldsymbol{x}$, can just be dropped.

We will write the reduced set of equations in the form

$$\hat{\tilde{M}} \cdot \tilde{\boldsymbol{x}} = \boldsymbol{y}^I \tag{10.18}$$

The vector $\tilde{\boldsymbol{x}}$ has less elements than $\boldsymbol{x}$, let us call this number N_R. The number of equations, however, is the same as that of the original set (10.11), namely N_M.

In case that only some of the symmetry relations are considered absolutely valid only these relations can be used to reduce the number of unknown correlations. The remaining relations can just be added to the system (10.13). Of course, for these equations the respective y_i are zero. For simplicity, equation (10.13) will be considered to comprise this case.

The most general way to calculate the minimum of $Q(\tilde{\boldsymbol{x}})$ under consideration of the symmetry relations (10.11) is to introduce particular weight factors, which take care of the fact, that the measured quantities y_i have different reliabilities. The theory requires that these factors are the reciprocal of the variance of the y_i. The size of the weight factors for those symmetry relations which are considered not strictly valid cannot be derived from theoretical considerations but must rather be sought in a trial and error manner.

The resulting form of the function $Q(\tilde{\boldsymbol{x}})$ will then be

$$Q(\tilde{\boldsymbol{x}}) = \sum_i w_i^2 \left(\sum_j M_{ij}\tilde{x}_i - y_j \right)^2 . \qquad (10.19)$$

With the introduction of the diagonal weight matrix $\hat{W}$:

$$W_{ij} = w_i^2 \cdot \delta_{ij} \text{ with } \delta_{ij} = \begin{cases} 1 & \text{for } i = j \\ 0 & \text{for } i \neq j \end{cases} \qquad (10.20)$$

we can write

$$Q(\tilde{\boldsymbol{x}}) = (\tilde{M} \cdot \tilde{\boldsymbol{x}} - \boldsymbol{y})^T \cdot \hat{W} \cdot (\tilde{M} \cdot \boldsymbol{x} - \boldsymbol{y}). \qquad (10.21)$$

For the case where no weight factors are included the matrix $\hat{W}$ is just a unit matrix.

At the minimum of this expression, all derivatives with respect to the components of $\tilde{\boldsymbol{x}}$ vanish. Consequently, it is required that

$$\frac{dQ(\tilde{\boldsymbol{x}})}{d\tilde{x}_i} = 0 \text{ for } i = 1 ... N_R. \qquad (10.22)$$

It is not difficult to calculate the derivatives which lead to the following set of equations:

$$\hat{\tilde{M}}^T \cdot \tilde{W} \cdot \hat{\tilde{M}} \cdot \tilde{\boldsymbol{x}} = \hat{\tilde{M}}^T \cdot \tilde{W} \cdot \boldsymbol{y} \qquad (10.23)$$

This is a set of inhomogeneous equations with N_R unknown $\tilde{x}_i$ and the same number of equations. This follows from the fact that the matrix $\hat{\tilde{M}}^T \cdot \tilde{W} \cdot \hat{\tilde{M}}$ is a square matrix, which can easily be proved: Let us write

$$\hat{A} = \hat{\tilde{M}}^T \cdot \tilde{W} \cdot \hat{\tilde{M}} \qquad (10.24)$$

An arbitrary element A_{ij} of $\hat{A}$ is

$$A_{ij} = \sum_{k,l} \tilde{M}_{ik}^T \cdot W_{kl} \cdot \tilde{M}_{lj} = \sum_k \tilde{M}_{ik}^T \cdot W_{kk} \cdot \tilde{M}_{kj} \qquad (10.25)$$

since $\hat{W}$ is a diagonal matrix. We now write down in an analogous way the general element of the transposed matrix $\hat{A}^T$:

$$\begin{aligned} A_{ij}^T = A_{ji} &= \sum_k \tilde{M}_{jk}^T \cdot W_{kk} \cdot \tilde{M}_{ki} && (10.26) \\ &= \sum_k \tilde{M}_{kj} \cdot W_{kk} \cdot \tilde{M}_{ik}^T \\ &= \sum_k \tilde{M}_{ik}^T \cdot W_{kk} \cdot \tilde{M}_{kj} \\ &= A_{ij} && (10.27) \end{aligned}$$

This proves that the matrix $\hat{A}$ is symmetric and, therefore, the system (10.23) of linear equations consists of the same number of equations as there are unknowns.

The equation (10.23) has a unique solution if the determinant of $\hat{A}$ does *not* vanish. In that case all correlations $\tilde{x}_i$ can be calculated. If the determinant of the matrix $\hat{M}$ vanishes, one cannot determine all correlations separately, and one or more linear combinations remain, which cannot be resolved.

10.2 Experiments and Results

The experimental set-up, which enabled the measurement of the data presented below, is shown schematically in Fig. 10.1.

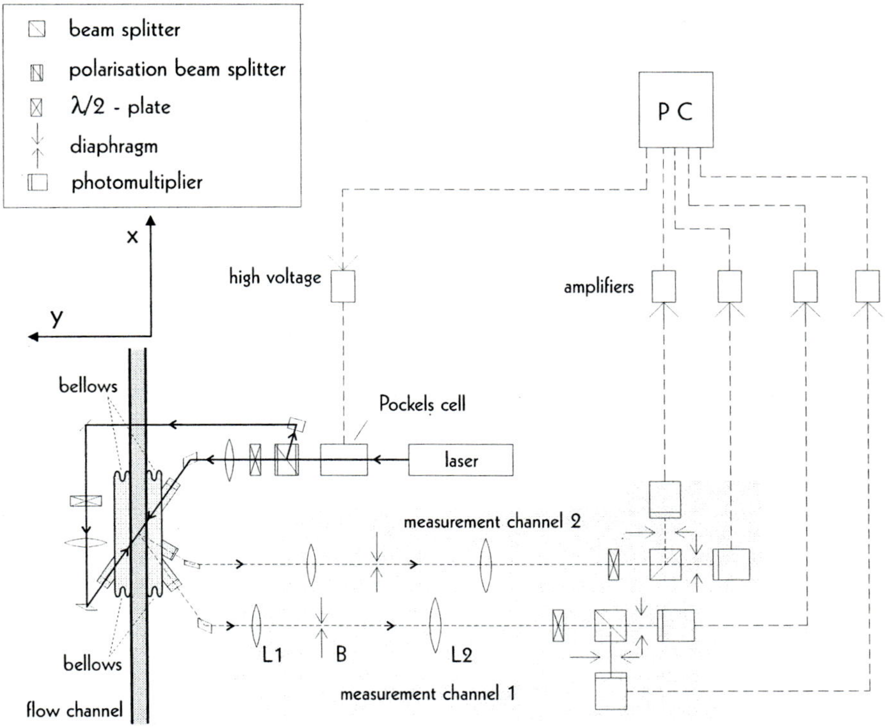

Fig. 10.1. Schematic representation of the experimental set-up to measure correlations of velocity gradients. Explanation in the text.

Light of a 1-W argon laser first passes through a Pockels cell which is controlled by the PC. Depending on the voltage applied to the Pockels cell, the light leaves the cell in either of two orthogonal linear polarization states. A polarizing beam splitter causes two different light paths for the two polarizations, leading to the opposite beam directions in the scattering volume. In each of the light paths a lambda half plate is used to rotate the plane of polarization, such that the light scattered into the measurement channels causes equal intensity in both channels. Of the measurement channels only the optic axes are drawn; the measurement channels are identical to the one shown in figure 5.3. The flow channel is described in Chap. 7.

To facilitate measurements of the y-dependence of gradient correlations, the whole optical set-up, apart from the illuminating laser, was built on a movable plate which could be traversed in the y-direction by use of high precision screws. Two water filled cuvettes with rubber bellows were attached to the side walls of the flow channel with the glass windows fixed to the

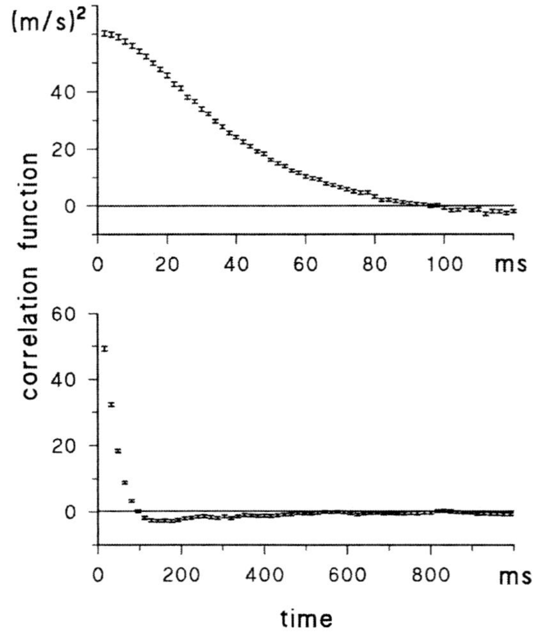

Fig. 10.2. Example of a measured auto correlation function of the pattern velocity. Note the different time scales!

movable plate. By this means a shift of the movable plate in the y-direction only changes the optical light path with respect to the flow channel, i.e. the location of the scattering volume within the flow changes, but no parameters describing the light path in the system of the moving plate will change. Therefore, no further adjustment is required if the position of the scattering volume is changed by shifting the movable plate.

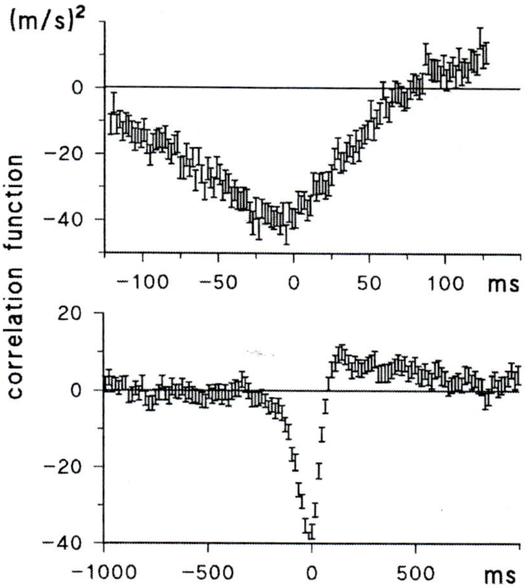

Fig. 10.3. Example of a measured cross-correlation function of different velocities of the pattern of the scattered light. Note the different time scales!

The pulses of the photomultiplier are transformed to normalized pulses after they passed through amplifiers and discriminators. They are then counted by self-built electronics. The count numbers are transferred to the PC after the end of each sample time. The PC calculates the pattern velocities and from them the gradient components and their correlation functions on the basis of equation (10.4).

Examples of gradient correlations, as calculated directly from the count rates fed to the PC, are shown in Figs. 10.2 and 10.3.

The proper correlation functions, i.e. those of components in the system of the flow, were calculated from 24 measured correlation functions. The relationships from Sect. 10.1.2 were used for the necessary transformations. Examples of the obtained correlation functions are shown in the Figs. 10.4 and 10.5.

For some quantities of interest in fluid dynamics only correlations of gradient values are of interest, rather than the actual temporal correlation functions. These correlations are simply the values in the zeroth channel of the corresponding correlation function, i.e. $C_{ij}(0)$. In Figs. 10.6 and 10.7, examples of variances of gradient correlations are shown. Although such quantities must always be positive by definition, negative values can be seen especially for small wall distances. This behaviour is caused by the fact that these values are calculated from measured data with a certain amount of error.

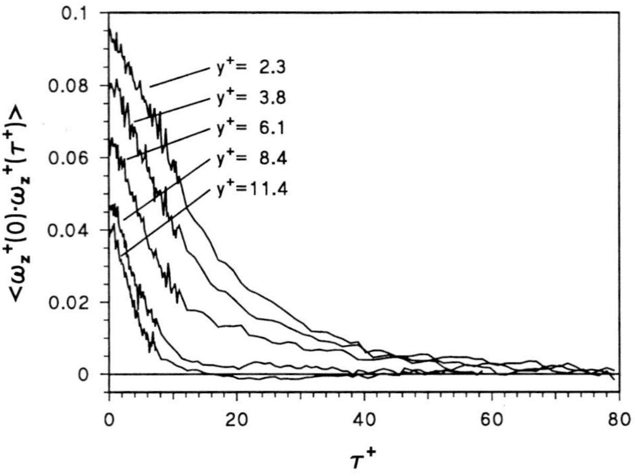

Fig. 10.4. Auto-correlation functions of the z-component of rot($\boldsymbol{u}$), (denoted by ω_z), for different wall distances.

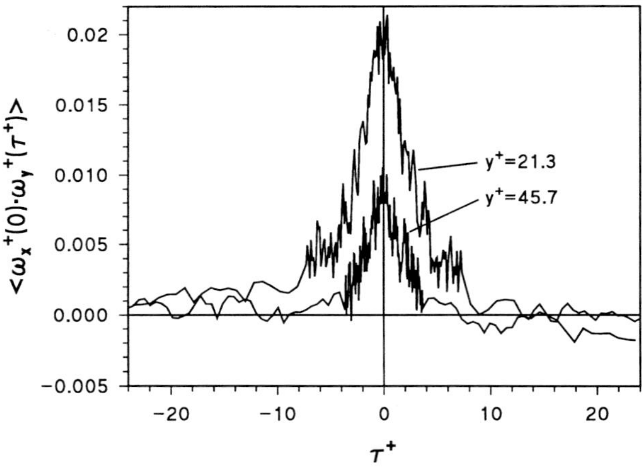

Fig. 10.5. The cross-correlation functions of the x- and y-components of rot($\boldsymbol{u}$) (denoted by ω_x and ω_y, respectively) for different wall distances.

In Fig. 10.7, errors of very different magnitudes for the different variances can be observed. This is a result of the very different coefficients which appear in the expressions that connect the measured correlations with the 'proper' correlations. Therefore, proper correlations, which appear in none of the relations with a large coefficient, cannot be calculated with good precision. If, therefore, particular gradient correlations are to be measured one has to choose scattering geometries where the corresponding coefficients are large, i.e. close to 1.

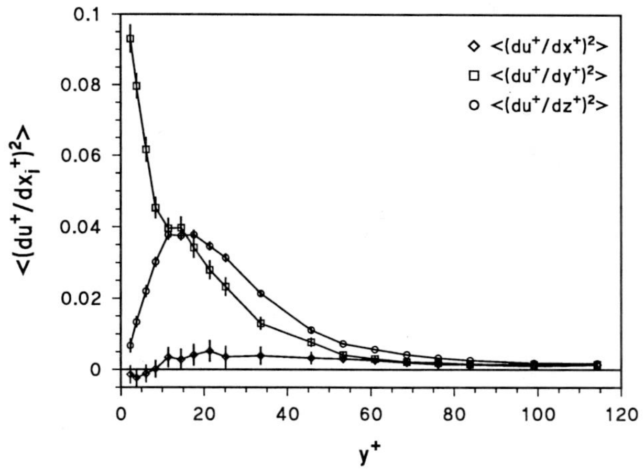

Fig. 10.6. Variances of the gradient component du/dx$_j$ as a function of the wall distance.

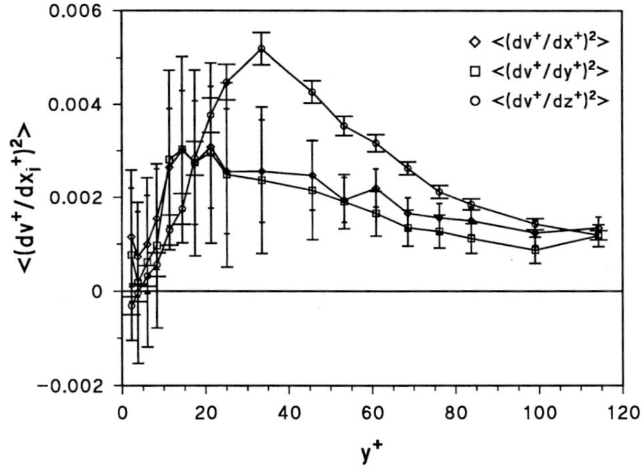

Fig. 10.7. Variances of the gradient component dv/dx$_j$ as a function of the wall distance.

Finally, Figs. 10.8 and 10.9 show the enstrophy and the turbulent energy dissipation as functions of the wall distance.

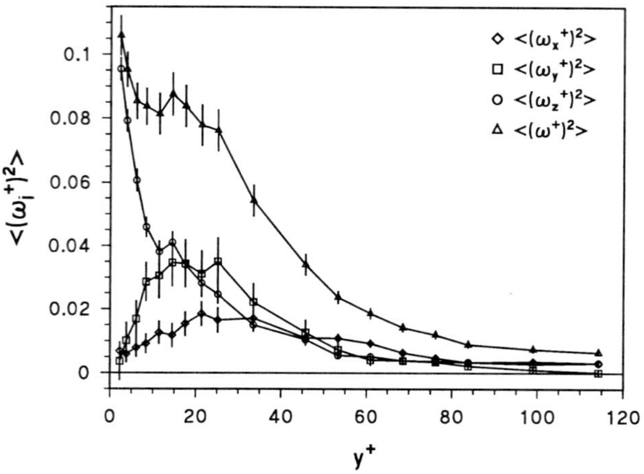

Fig. 10.8. The enstrophy and the components of vorticity as a function of the wall distance.

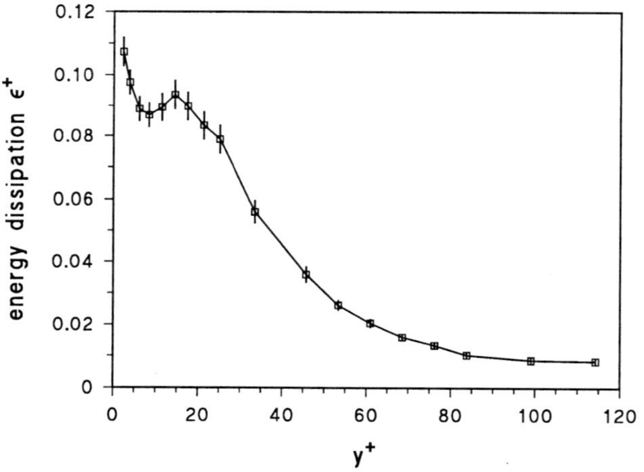

Fig. 10.9. The turbulent energy dissipation as a function of the wall distance.

11. Outlook

For a general assessment of the methods described in this book let us first consider whether the basic physical effect is suited to serve as a foundation of reliable and versatile methods to measure velocity gradients. It is not difficult to list the advantages: Measurement schemes based on it are:

- non-invasive
- gauge-free
- direct.

A further advantage over other optical methods concerns the necessary optical inhomogeneities: FRS needs no seeding with scattering particles at all, and no slip between inhomogeneity and fluid can occur. The other methods can use sufficiently small particles to render slip negligible.

If the shortcomings are to be discussed one has first to consider that the methods described are *special* applications of the basic effect. Of course, any of the mentioned methods has its shortcomings, but that are to a great deal different ones with the different methods. For instance, DSS is sensitive to stray light whereas DPSS and FRS are not, FRS requires a special tuning of the laser power to the absorption constant of the fluid, in DPSS the scattering volume cannot be made as small as in DSS and CCS. So, depending on the experimental demands one is not forced to accept certain disadvantages of one particular method. Furthermore, it is worth while stressing that so far only very few research groups were engaged in the development and application of the mentioned effect. Consequently, there was not very much activity so far to develop alternative methods to the ones described here or to introduce major improvements.

But even with the methods described here and their shortcomings interesting quantities can be measured. Especially for measurements of wall gradients the experimental requirements are comparatively little as shown in chapter 5.9.2. So this type of application seems most likely to be of interest for the applied fluid flow researcher and also in rheology.

The measurement of a shear or a vorticity component requires two optical set-ups - two illuminating laser beams and two detection channels - and a device to switch the two beams on and off for the two velocity gradient components which they are composed of (c.f. chapter 5.9.1). The demands on the whole set-up are therefore a lot higher and at least with the methods described presumably only scientists doing basic research will consider an implementation.

As discussed in this chapter, direct measurements of on-diagonal elements of the velocity gradient tensor are not possible, only linear combinations of on- and off-diagonal elements can be measured with one set-up. To measure *all* components is of course really quite complex, but nevertheless, it was shown that the difficulties can be surmounted.

So far these methods have only been applied to fluids. But one should not deduce from this fact that they cannot be successfully applied to gas flows, too, with the exception of FRS. Of course, there must be enough particles in the chosen scattering volume. So far optical measurements in gases as LDA or imaging methods (see chapter 2, section 2.3) were performed which must take care to have a particle density which is not too high. But for the methods described here there is rather the problem of the particle concentration to be too low. Of course, pattern motion occurs even if there are only two particles in the scattering volume. So very likely DPSS works also with gaseous flows. The situation might be somewhat different with DSS: It is not obvious whether large fluctuations in the particle number has no influence on the data. Computer simulations by the author suggest that the DSS-method should give reliable results independent of the average particle number as far as average velocity gradient values are concerned, but this must be checked experimentally. The measurement of velocity gradient correlations poses a problem as discussed in the paragraph following equation 10.1: The fluctuation of the light intensity will be large if the average number of particles within the scattering volume is small, and therefore correlations between the intensities of light scattered in different directions occurs which are not caused by the velocity gradient. To obtain a sufficiently high intensity of the scattered light is also not simple with gases and high laser power might lead to accelerations of the particles. As a result, for successful applications of the method to gas flows there are still a number of questions to be answered.

Last but not least it should be mentioned that the methods to measure pattern velocities can also be used to measure flow velocities. Only the imaging in the detection channel has to be changed: Instead of placing the detector in the focal plane of the optical system one has to image the particles in the scattering volume onto the detector. This method has already successfully been applied to measure the velocity of a fluid in a rotating cuvette [24]. The same optical set-up was used for measuring both, velocity and velocity gradient. The 'switching' between the two modes was simply done by shifting one lens.

The possibility to measure velocity gradients and velocities with essentially the same set-up allows to determine experimentally velocity-velocity gradient correlations, which appear in many equations in fluid dynamics (see for instance equation (2.15)).

Bibliography

1. P.Vukoslavčević, J.M.Wallace und J.-L.Balint: 'The velocity and vorticity vector fields of a turbulent boundary layer. Part 1: Simultaneous measurement by hot wire anemometry, Part 2: Statistical properties', J.Fluid.Mech. **228**(1991) 25-86
2. N.A. Fomin: 'Speckle Photography for Fluid Mechanics Measurements', Springer-Verlag Berlin-Heidelberg (1998)
3. T. Urushihara, C.D. Meinhart, and R.J. Adrian: 'Investigation of the logarithmic layer in pipe flow using particle image velocimetry', Near-wall turbulent flows: Proceedings of the International Conference on Near Wall Turbulent Flows, Tempe, Arizona USA (1993)434 - 446
4. J.H. Agüí and J. Andreopoulos: 'Development of a new laser vorticity probe - LAVOR' Annual Symp. on Laser Anemometry, ASME Fluid Engr. Meet., June 19-23 Lake Tahoe, California USA (1994)
5. Y.Aizu, T. Ushizaka, and T. Asakura: 'Measurements of the Velocity Gradient using a Laser Doppler Phenomenon', Appl. Phys. B **36** (1985)155 - 161
6. M.V. Ötügen, W.Su, and G.Papadopoulos: 'A new laser-based method for strain rate and vorticity measurements', Meas. Sci. Technol. **9** (1998) 267 - 274
7. M.B.Frish und W.W.Webb: 'Direct measurement of vorticity by optical probe', J.Fluid.Mech **107**(1981) 173-200
8. A.A: Naqwi and W. C. Reynolds: 'Measurement of turbulent wall velocity gradients using cylindrical waves of laser light', Experiments in Fluids **10** (1991)257 - 266
9. G.G. Fuller, J.M. Rallison, R.L. Schmidt, and L.G.Leal: 'The measurement of velocity gradients in laminar flow by homodyne light-scattering spectroscopy', J. Fluid Mech. **100** part 3, (1980)555 - 575
10. W.Staude: 'The velocity of random time-dependent patterns and its experimental determination', J.Phys. D **29** (1996) 307-314
11. C.Keveloh und W.Staude: 'The measurement of velocity gradients in laminar and turbulent flow', J. Phys. D: Appl. Phys. **21**, (1988)237-245
12. H.Kogelnik und T.Li: 'Laser Beams and Resonators', Applied Optics **5**(1966) 1550-1567
13. H.C. van de Hulst: 'Light scattering by small particles', Dover Publications New York 1981
14. M.Fermigier, P.Jenffer, J.C.Charmet und E.Guyon: 'A non-perturbative anemometric and flow visualisation technique', J.Physique Lett. **41** (1980) L-519 - L-521
15. See for instance N.Ahmed: 'Fluid Mechanics' (San Jose, Ca: Engenieering Press 1987) p 326
16. P.G. de Gennes: 'Principe de nouvelle mesure sur les écoulements par échauffements optique localisés', J.Physique Lett. **38**(1977) L-1 - L-3

17. M.Cloitre und E.Guyon: 'Forced Rayleigh scattering in turbulent plane Poiseuille flows', J.Fluid Mech. **164** (1986) 217-236
18. J.H. Agüí and L. Hesselink: 'Application of holography to the measurment of velocity gradients in fluid flows sensitized with photochromic dyes', Physics of Fluids **A2** (1990) 688-699
19. U.Schmidt: 'Ermittlung von Reynoldsstreß, Energieproduktions- und Dissipationsraten in turbulenter Kanalströmung mittels Laser-Speckle-Strophometrie', Disseration, Bremen 1990
20. C.Keveloh: 'Meßverfahren zur Messung von Geschwindigkeitsgradienten und deren Verteilungen in laminarer und turbulenter Strömung', Dissertation, Bremen 1988
21. H.Kriegs: 'Die Doppelpulse-Speckle-Strophometrie - ein optisches Verfahren zur Messung der Momentanwerte von Geschwindigkeitsgradienten in turbulenten Strömungen', Dissertation, Bremen 1993
22. H.Kriegs und W.Staude: 'A laser pulse technique for the measurement of time-resolved velocity gradients in fluid flow', Meas. Sci. Technol **6** (1995) 653-662
23. H.Breyer und W.Staude: 'The differential speckle velocimetry - a method to measure the velocity of random patterns', Opt. Commun. **73**, 422(1989)
24. H. Breyer 'Die differentielle Speckle-Anemometrie - ein neues fluiddynamisches Meßverfahren', Dissertation, Bremen 1991
25. H.Kriegs, R.Schulz und W.Staude: 'The measurement of velocity gradients in fluid flow by laser light scattering, Part2: Statistical properties of gradients', Exp. in Fluids **15**, (1993)240-246
26. R.Schulz and W.Staude : 'Determination of the velocity gradient correlation in a turbulent channel flow by laser-light scattering', Journal of Physics D: Applied Physics **31**(1998) 3066-3081
27. H.Tennekes und J.L.Lumley: 'A first course in turbulence', Cambridge Massachusetts, 1972, MIT press

Subject Index

absorption length 89
achromatic lens 86
aperture function 60
average value
– definition 24
– ensemble 26
– spatial 28
– temporal 25

beam waist 83
boiling *see* speckle pattern
Brownian motion 23, 27, 77

channel flow 110
coherence 67
– area 68
– volume *see* q-volume of coherence
coherent structures 9
composed rotations 46
correlation function 35
– auto- 35
– cross- 35
correlation time 39

dark current 91
dead time 142
dimensionless variables 110
direct imaging 13
Doppler effect 15

electric field 52
enstrophy 111
ergodic process 27
ergodicity *see* ergodic process
Ewald sphere 61, 68

frequency of a light wave 52

Gaussian beam 84

homogeneity 28
homogeneity of a stochastic process 33
homogeneity volume 12
homogeneous process 28

illuminating laser beam 83
initial phase of a wave 52
isotropic process 28

kinematic viscosity 6

Laser Doppler anemometry 15
Laser Doppler beat 17
LDA *see* Laser Doppler anemometry
light intensity 55
light wave
– amplitude of 12, 53, 54
– complex amplitude of 53
– definition of amplitude 52
light-sheet technique 13
logarithmic wall law 111

matrix
– inverse 45
– transpose 45
– unit 45
measurement volume 12

Navier-Stokes equation 5
noise 24

phase of a wave 53
phase surface 72
phase velocity 53
photon number 24
plane light wave 83
polarization vector of a light wave 52
probability
– conditional 31
– density 28
– joint 30
– – order of 31

q-volume of coherence 63

reflection at a mirror 99
reflection matrix 99, 100
refractive index 53
Reynolds decomposition 6
Reynolds number 110
Reynolds stress 7
rotation
– of a co-ordinate system 45
– of a tensor 48
– of a vector 45
rotation matrix 45

scattering factor 57
scattering length 89
scattering vector 60
scattering volume 12, 23, 60
seeding 12
shear stress 5
shot noise 24
slip 12, 13, 173
spatial aperture function 87
speckle motion
– delusive 79
– oblique 75
speckle pattern 63, 64
– boiling of 40, 74, 77
– spatial correlation of 63, 69
– temporal change 71
– velocity of 74
speckle size 63
spherical aberration 86
spherical wave 54
stationarity of a stochastic process 33
stationary process 26
stochastic process 24, 26–28
stochastic variable 29
stray light
– definition 91
– coherent 91
– incoherent 91
system of the flow 101, 102
system of the light 100, 101

TEM_{oo}-mode 83
tensor
– definition 47
– transformation of 43
– symmetric part of 50
– antisymmetric part of 50
– functions of a 48
transformation of variables
– in probability densities 32
turbulence 6
turbulent energy dissipation 111

velocity gradient tensor
– definition 49
– measurement of
– – off-diagonal elements 103
– – on-diagonal elements 104
– symmetry of 50
visco-elasticity 5
vorticity 8, 14, 19, 50
– measurement of 104, 144, 173

wall gradient 9, 92
– measurement of 19, 105
wave vector 52
wavelength 52
weight functions 148

Lecture Notes in Physics

For information about Vols. 1–542
please contact your bookseller or Springer-Verlag

Vol. 543: H. Gausterer, H. Grosse, L. Pittner (Eds.), Geometry and Quantum Physics. Proceedings, 1999. VIII, 408 pages. 2000.

Vol. 544: T. Brandes (Ed.), Low-Dimensional Systems. Interactions and Transport Properties. Proceedings, 1999. VIII, 219 pages. 2000

Vol. 545: J. Klamut, B. W. Veal, B. M. Dabrowski, P. W. Klamut, M. Kazimierski (Eds.), New Developments in High-Temperature Superconductivity. Proceedings, 1998. VIII, 275 pages. 2000.

Vol. 546: G. Grindhammer, B. A. Kniehl, G. Kramer (Eds.), New Trends in HERA Physics 1999. Proceedings, 1999. XIV, 460 pages. 2000.

Vol. 547: D. Reguera, G. Platero, L. L. Bonilla, J. M. Rubí(Eds.), Statistical and Dynamical Aspects of Mesoscopic Systems. Proceedings, 1999. XII, 357 pages. 2000.

Vol. 548: D. Lemke, M. Stickel, K. Wilke (Eds.), ISO Surveys of a Dusty Universe. Proceedings, 1999. XIV, 432 pages. 2000.

Vol. 549: C. Egbers, G. Pfister (Eds.), Physics of Rotating Fluids. Selected Topics, 1999. XVIII, 437 pages. 2000.

Vol. 550: M. Planat (Ed.), Noise, Oscillators and Algebraic Randomness. Proceedings, 1999. VIII, 417 pages. 2000.

Vol. 551: B. Brogliato (Ed.), Impacts in Mechanical Systems. Analysis and Modelling. Lectures, 1999. IX, 273 pages. 2000.

Vol. 552: Z. Chen, R. E. Ewing, Z.-C. Shi (Eds.), Numerical Treatment of Multiphase Flows in Porous Media. Proceedings, 1999. XXI, 445 pages. 2000.

Vol. 553: J.-P. Rozelot, L. Klein, J.-C. Vial Eds.), Transport of Energy Conversion in the Heliosphere. Proceedings, 1998. IX, 214 pages. 2000.

Vol. 554: K. R. Mecke, D. Stoyan (Eds.), Statistical Physics and Spatial Statistics. The Art of Analyzing and Modeling Spatial Structures and Pattern Formation. Proceedings, 1999. XII, 415 pages. 2000.

Vol. 555: A. Maurel, P. Petitjeans (Eds.), Vortex Structure and Dynamics. Proceedings, 1999. XII, 319 pages. 2000.

Vol. 556: D. Page, J. G. Hirsch (Eds.), From the Sun to the Great Attractor. X, 330 pages. 2000.

Vol. 557: J. A. Freund, T. Pöschel (Eds.), Stochastic Processes in Physics, Chemistry, and Biology. X, 330 pages. 2000.

Vol. 558: P. Breitenlohner, D. Maison (Eds.), Quantum Field Theory. Proceedings, 1998. VIII, 323 pages. 2000

Vol. 559: H.-P. Breuer, F. Petruccione (Eds.), Relativistic Quantum Measurement and Decoherence. Proceedings, 1999. X, 140 pages. 2000.

Vol. 560: S. Abe, Y. Okamoto (Eds.), Nonextensive Statistical Mechanics and Its Applications. IX, 272 pages. 2001.

Vol. 561: H. J. Carmichael, R. J. Glauber, M. O. Scully (Eds.), Directions in Quantum Optics. XVII, 369 pages. 2001.

Vol. 562: C. Lämmerzahl, C. W. F. Everitt, F. W. Hehl (Eds.), Gyros, Clocks, Interferometers...: Testing Relativistic Gravity in Space. XVII,507 pages. 2001.

Vol. 563: F. C. Lázaro, M. J. Arévalo (Eds.), Binary Stars. Selected Topics on Observations and Physical Processes. 1999.IX, 327 pages. 2001.

Vol. 564: T. Pöschel, S. Luding (Eds.), Granular Gases. VIII, 457 pages. 2001.

Vol. 565: E. Beaurepaire, F. Scheurer, G. Krill, J.-P. Kappler (Eds.), Magnetism and Synchrotron Radiation. XIV, 388 pages. 2001.

Vol. 566: J. L. Lumley (Ed.), Fluid Mechanics and the Environment: Dynamical Approaches. VIII, 412 pages. 2001.

Vol. 567: D. Reguera, L. L. Bonilla, J. M. Rubí (Eds.), Coherent Structures in Complex Systems. IX, 465 pages. 2001.

Vol. 568: P. A. Vermeer, S. Diebels, W. Ehlers, H. J. Herrmann, S. Luding, E. Ramm (Eds.), Continuous and Discontinuous Modelling of Cohesive-Frictional Materials. XIV, 307 pages. 2001.

Vol. 569: M. Ziese, M. J. Thornton (Eds.), Spin Electronics. XVII, 493 pages. 2001.

Vol. 570: S. G. Karshenboim, F. S. Pavone, F. Bassani, M. Inguscio, T. W. Hänsch (Eds.), The Hydrogen Atom: Precision Physics of Simple Atomic Systems. XXIII, 293 pages. 2001.

Vol. 571: C. F. Barenghi, R. J. Donnelly, W. F. Vinen (Eds.), Quantized Vortex Dynamics and Superfluid Turbulence. XXII, 455 pages. 2001.

Vol. 572: H. Latal, W. Schweiger (Eds.), Methods of Quantization. XI, 224 pages. 2001.

Vol. 573: H. M. J. Boffin, D. Steeghs, J. Cuypers (Eds.), Astrotomography. XX, 432 pages. 2001.

Vol. 574: J. Bricmont, D. Dürr, M. C. Galavotti, G. Ghirardi, F. Petruccione, N. Zanghi (Eds.), Chance in Physics. XI, 288 pages. 2001.

Vol. 575: M. Orszag, J. C. Retamal (Eds.), Modern Challenges in Quantum Optics. XXIII, 405 pages. 2001.

Vol. 577: N. Thomas, I. P. Williams (Eds.), Solar and Extra-Solar Planetary Systems. Approx. X, 266 pages. 2001.

Vol. 578: D. Blaschke, N. K. Glendenning, A. Sedrakian (Eds.), Physics of Neutron Star Interiors. XI, 511 pages. 2001.

Vol. 579: R. Haug, H. Schoeller (Eds.), Interacting Electrons in Nanostructures. X, 227 pages. 2001.

Vol. 580: K. Baberschke, M. Donath, W. Nolting (Eds.), Band-Ferromagnetism: Ground-State and Finite-Temperature Phenomena.IX, 394 pages. 2001.

Vol.581: J. M. Arias, M. Lozano (Eds.), An Advanced Course in Modern Nuclear Physics. XI, 362 pages. 2001.

Monographs

For information about Vols. 1–28
please contact your bookseller or Springer-Verlag

Vol. m 29: C. Bendjaballah, Introduction to Photon Communication. VII, 193 pages. 1995.

Vol. m 30: A. J. Greer, W. J. Kossler, Low Magnetic Fields in Anisotropic Superconductors. VII, 161 pages. 1995.

Vol. m 31 (Corr. Second Printing): P. Busch, M. Grabowski, P.J. Lahti, Operational Quantum Physics. XII, 230 pages. 1997.

Vol. m 32: L. de Broglie, Diverses questions de mécanique et de thermodynamique classiques et relativistes. XII, 198 pages. 1995.

Vol. m 33: R. Alkofer, H. Reinhardt, Chiral Quark Dynamics. VIII, 115 pages. 1995.

Vol. m 34: R. Jost, Das Märchen vom Elfenbeinernen Turm. VIII, 286 pages. 1995.

Vol. m 35: E. Elizalde, Ten Physical Applications of Spectral Zeta Functions. XIV, 224 pages. 1995.

Vol. m 36: G. Dunne, Self-Dual Chern-Simons Theories. X, 217 pages. 1995.

Vol. m 37: S. Childress, A.D. Gilbert, Stretch, Twist, Fold: The Fast Dynamo. XI, 406 pages. 1995.

Vol. m 38: J. González, M. A. Martín-Delgado, G. Sierra, A. H. Vozmediano, Quantum Electron Liquids and High-Tc Superconductivity. X, 299 pages. 1995.

Vol. m 39: L. Pittner, Algebraic Foundations of Non-Com-mutative Differential Geometry and Quantum Groups. XII, 469 pages. 1996.

Vol. m 40: H.-J. Borchers, Translation Group and Particle Representations in Quantum Field Theory. VII, 131 pages. 1996.

Vol. m 41: B. K. Chakrabarti, A. Dutta, P. Sen, Quantum Ising Phases and Transitions in Transverse Ising Models. X, 204 pages. 1996.

Vol. m 42: P. Bouwknegt, J. McCarthy, K. Pilch, The W3 Algebra. Modules, Semi-infinite Cohomology and BV Algebras. XI, 204 pages. 1996.

Vol. m 43: M. Schottenloher, A Mathematical Introduction to Conformal Field Theory. VIII, 142 pages. 1997.

Vol. m 44: A. Bach, Indistinguishable Classical Particles. VIII, 157 pages. 1997.

Vol. m 45: M. Ferrari, V. T. Granik, A. Imam, J. C. Nadeau (Eds.), Advances in Doublet Mechanics. XVI, 214 pages. 1997.

Vol. m 46: M. Camenzind, Les noyaux actifs de galaxies. XVIII, 218 pages. 1997.

Vol. m 47: L. M. Zubov, Nonlinear Theory of Dislocations and Disclinations in Elastic Body. VI, 205 pages. 1997.

Vol. m 48: P. Kopietz, Bosonization of Interacting Fermions in Arbitrary Dimensions. XII, 259 pages. 1997.

Vol. m 49: M. Zak, J. B. Zbilut, R. E. Meyers, From Instability to Intelligence. Complexity and Predictability in Nonlinear Dynamics. XIV, 552 pages. 1997.

Vol. m 50: J. Ambjørn, M. Carfora, A. Marzuoli, The Geometry of Dynamical Triangulations. VI, 197 pages. 1997.

Vol. m 51: G. Landi, An Introduction to Noncommutative Spaces and Their Geometries. XI, 200 pages. 1997.

Vol. m 52: M. Hénon, Generating Families in the Restricted Three-Body Problem. XI, 278 pages. 1997.

Vol. m 53: M. Gad-el-Hak, A. Pollard, J.-P. Bonnet (Eds.), Flow Control. Fundamentals and Practices. XII, 527 pages. 1998.

Vol. m 54: Y. Suzuki, K. Varga, Stochastic Variational Approach to Quantum-Mechanical Few-Body Problems. XIV, 324 pages. 1998.

Vol. m 55: F. Busse, S. C. Müller, Evolution of Spontaneous Structures in Dissipative Continuous Systems. X, 559 pages. 1998.

Vol. m 56: R. Haussmann, Self-consistent Quantum Field Theory and Bosonization for Strongly Correlated Electron Systems. VIII, 173 pages. 1999.

Vol. m 57: G. Cicogna, G. Gaeta, Symmetry and Perturbation Theory in Nonlinear Dynamics. XI, 208 pages. 1999.

Vol. m 58: J. Daillant, A. Gibaud (Eds.), X-Ray and Neutron Reflectivity: Principles and Applications. XVIII, 331 pages. 1999.

Vol. m 59: M. Kriele, Spacetime. Foundations of General Relativity and Differential Geometry. XV, 432 pages. 1999.

Vol. m 60: J. T. Londergan, J. P. Carini, D. P. Murdock, Binding and Scattering in Two-Dimensional Systems. Applications to Quantum Wires, Waveguides and Photonic Crystals. X, 222 pages. 1999.

Vol. m 61: V. Perlick, Ray Optics, Fermat's Principle, and Applications to General Relativity. X, 220 pages. 2000.

Vol. m 62: J. Berger, J. Rubinstein, Connectivity and Superconductivity. XI, 246 pages. 2000.

Vol. m 63: R. J. Szabo, Ray Optics, Equivariant Cohomology and Localization of Path Integrals. XII, 315 pages. 2000.

Vol. m 64: I. G. Avramidi, Heat Kernel and Quantum Gravity. X, 143 pages. 2000.

Vol. m 65: M. Hénon, Generating Families in the Restricted Three-Body Problem. Quantitative Study of Bifurcations. XII, 301 pages. 2001.

Vol. m 66: F. Calogero, Classical Many-Body Problems Amenable to Exact Treatments. XIX, 749 pages. 2001.

Vol. m 67: A. S. Holevo, Statistical Structure of Quantum Theory. IX, 159 pages. 2001.

Vol. m 68: N. Polonsky, Supersymmetry: Structure and Phenomena. Extensions of the Standard Model. XV, 169 pages. 2001.

Vol. m 69: W. Staude, Laser-Strophometry. High-Resolution Techniques for Velocity Gradient Measurements in Fluid Flows. XV, 178 pages. 2001